The Patchwork Human

The Patchwork Human

*Two Billion Years
of Evolution*

PETER LUYKX

McFarland & Company, Inc., Publishers
Jefferson, North Carolina

Illustrations are by Kate Baldwin unless otherwise noted.

ISBN (print) 978-1-4766-8612-7
ISBN (ebook) 978-1-4766-4565-0

LIBRARY OF CONGRESS AND BRITISH LIBRARY
CATALOGUING DATA ARE AVAILABLE

Library of Congress Control Number 2022009215

Front cover image: "Woman in Red Blouse," by Chaim Soutine,
1919, Russian French Expressionist painting,
oil on canvas (Everett Collection/Shutterstock)

Printed in the United States of America

*McFarland & Company, Inc., Publishers
Box 611, Jefferson, North Carolina 28640
www.mcfarlandpub.com*

For Jan, Aurolyn, and Diana,
the women who keep me human

Acknowledgments

I am grateful to Nancy Hardin and John Rolfe Gardiner for their early encouragement. I had many valuable early discussions with Floyd Rosen, Keith Waddington, and Sawsan Khuri, who provided useful comments on several sections of the book. Tara Kai early on nudged me out of a rut and onto a better path. Sharon Buydens proofread the section on altruism. Thanks go to the memory of the late Catherine Fisher Lawler, who in spite of her protestations turned out to be the Perfect Reader after all. I also thank Peter Lamar, Efrain Zabala, and Kelley Flannery Rowan for useful discussions especially on the subject of religion and evolution.

I thank my many colleagues in the Biology Department at the University of Miami who opened my eyes to an evolutionary perspective after all my early training in the purely mechanistic view of life. Students in my classes in human genetics for non-scientists provided much of the stimulus and motivation for this book. Members of the classes I taught at the Osher Lifelong Learning Institute at the University of Miami gave me the incentive to explore areas of human evolution I might otherwise have neglected.

Harvey Siegel and Otávio Bueno, my friends in the Philosophy Department, provided useful background (and sometimes foreground) and sharpened my thinking on many points.

Finally, for their critical and insightful comments, I am grateful to the members of the Non-Fiction Critique Group of the South Florida Writers Association. I have not followed their advice in all instances, but their comments made me re-think many ideas and express them more clearly. It would be pleasant to say that any errors and obscurities that remain are their fault; alas, such is not the case.

Table of Contents

Preface

Some years ago, as a new member of the biology faculty, I was in the hallway chatting with a friend, a scholar of human social interactions. As another colleague of ours passed by, I glanced at him, wordlessly raising my eyebrows in greeting. "Ah," said my friend, "the old primate eyebrow flash." For a fleeting moment this observation was unwelcome because it evoked my evolutionary origins and reminded me that I was trapped in an evolutionary cage, not free even to choose my own facial expressions.

Recently, somewhat older and wiser, I stopped in at a local coffeeshop. It was crowded, filled with strange-looking upright apes. They were relatively hairless except for the tops of their heads where the hair often grew long and luxurious, especially on the females, although not at all on some of the males. The males also tended to be slightly larger than the females. All the foreheads bulged. There was an incessant chatter back and forth among them, the point of which was difficult to discern. It seemed as if these apes had an inner compulsion to produce an almost continuous stream of complicated sounds, which tended to be accompanied by a variety of facial expressions and hand gesticulations.

There is value in adding this kind of perspective to how we view ourselves—from outside, as it were, instead of always just swimming thoughtlessly and without objectivity inside our own biology. We can live both subjectively and objectively. We don't have to choose between appreciating the beauty of a rainbow and analyzing it in terms of wavelength, reflection, and refraction. We can do both and know the rainbow—and ourselves—inside and out.

My graduate studies in biology were aimed at understanding *mechanisms*: how living organisms worked at the cellular and molecular levels. That focus continued through my first and into my second academic post. In the second, by osmosis, I absorbed from my new colleagues some additional perspective. Important questions can be asked not only about how things work but also about how they might have got that way in the first place—evolutionary questions.

Who isn't interested in that primal question, *Where did I come from?* It would be easy to couch an answer in scientific jargon, as if it were inscribed on a stone tablet at the top of some mountain that can only be reached with the help of advanced education and technical training and by way of the narrowest of pathways from the plains below, where most people live. I've always preferred living on the plains, not so isolated from everyone else. For me, as for most people, explanations are better when they are simple. Therefore, this book, while scientific in content, is as free of scientific jargon as possible. It is nevertheless based on "hard science"—not "hard" as in "difficult," but "hard" as in "solid," as the scientific articles and books listed at the ends of the chapters testify.

Both interests, mechanisms and origins, are combined in this book. It covers earlier beginnings—from early life on this planet—than do most other books about human evolution, which usually start only with the origin of our species.

The introduction orients the reader to the idea that the human body is a jumble of parts, some very old (evolutionarily speaking), some newer. The first chapter indulges in a semi-philosophical discussion about whether there is fundamentally anything more to being human than cells and molecules: the physicalist says no, the dualist says yes. The next four chapters deal with material that is the foundation of everything else: the roles played by our small parts (cells and molecules), how the body develops from a fertilized egg (embryology), how change occurs over long periods of time (evolution's process), and what the record actually shows about our evolution (evolution's pattern). Following chapters ask, *Why are we made of so many cells? Why are our two sides mirror images of each other? Why do we use sex for reproduction? What does it mean to be a mammal? Why have we lost most of our fur?* and other questions. Later evolutionary additions to the human frame include walking upright, having a big brain, being a highly social animal, having music, having language, being highly variable (genetically speaking) and having traits that vary with geography. The final chapter examines some possibilities for the future of our species.

The "Background and Further Reading" section at the end of the book gives a list of some of the important scientific papers on which the factual statements in each chapter rely; many of these (although not all) are rather technical. It also provides some additional readings that go a little deeper into the subjects covered.

Like any species, humans have hundreds or even thousands of traits one could wonder about. I've chosen a few interesting ones. The information and ideas I relate in this book stem from a long-term interest in human genetics and evolution, from my research in those areas, and from years of reading and teaching at the undergraduate, graduate, and adult education levels. I am grateful to the people at all those levels for asking questions; they have sharpened my own understanding. Answers to a few of those questions can be found on my website, www.genetics-q-and-a.com.

This book aims to explain at least some of our traits—how we social, sexy, language-obsessed, technological apes came to be the way we are. "To explain" is a metaphor; it comes from the Latin *ex-*, out, and *plānum*, flat or level ground. As Nature is given to us, it's all rough terrain. Scientific explanation tries to lay the world out flat so that we can see it more clearly and understand it. It aims to provide a picture that enables us to appreciate better the lay of the land—the nearby better-known hills that are accessible from the commonsense plains where we live and the distant mysterious mountain peaks we have yet to explore.

Figure 1. "Explanation": extending our familiarity with Nature's flat landscape where we live, looking toward the nearby hills where our knowledge is secure, and viewing the distant mountains of unexplored territory (after Utagawa Hiroshige, 1797–1858, "Evening view of a temple in the hills") (Metropolitan Museum of Art, Henry L. Phillips Collection, Bequest of Henry L. Phillips, 1939).

Introduction

The Patchwork Human, Old and New Parts

...putting new wine into old wineskins...
—Luke 5:37, King James Version

The topics in this book go back about two billion years. Life itself began about four billion years ago on our four-and-a-half-billion-year-old planet. We don't yet have a coherent picture of how that might have happened. The earliest life forms must have been simple cells, something like present-day bacteria, although no doubt even simpler. It took another couple of billion years for the more complex kind of cell we're made of to arise (eukaryotic, or "true nucleus" cells), about two billion years ago (see Chapter 2, Figure 8). That's a starting point early enough for this book.

My grandmother used to make elaborate quilts, sewing together pieces of old, worn-out clothing. Some of them would take years to complete. When an old pair of pants was ready to be thrown out, she would cut usable parts of it into squares and put them in a basket with other pieces she had collected. When she was in the mood, she would take them out and start stitching them together, putting her work aside when she ran out of material that accorded with her sense of design. When enough old shirts or dresses or jackets or other items had accumulated, she would stitch the new pieces onto her previous work. At some point she got the idea of making not quilts but clothing that could be worn (Figure 2).

Imagine that such clothing is not something we *wear* but represents something woven into the very fabric of our being. We *are* that clothing, stitched together from old and new pieces by evolution.

It might seem that we are all of a piece, all the parts working together seamlessly to make us human. We could imagine that at some time in the past we were assembled in our present form all in one sewing session: this part from one basket, that part from another, a zipper here, a few molecular

4

buttons there, and you have a human being.

However, this doesn't describe even metaphorically how the human body came into being. Instead—as with any animal or plant—in early versions of ourselves now long gone, small modifications were made one by one during the course of evolution, until over time there gradually emerged a version different in many details from the original. At every stage during the long transformation all the parts worked together, as they continue to do in the modern version. Some parts of the original remain more or less intact, while other parts have been transformed beyond recognition. The modern human being is a patchwork of old and new parts. Our patchwork anatomy is shown in Figure 3: some very old parts and (in clockwise order) successively newer and newer parts.

The mechanism of change, described in more detail in Chapter 4, is analogous to how a memo might be updated periodically to reflect changed circumstances.

Meeting: Room 219, 8:30 a.m.; no. of persons, 10; Ms. Miller presiding

Figure 2. Patchwork clothing made of pieces of old material, some old, some newer (model Tamara Bellis, unsplash. com).

Ms. Miller comes down with the flu, and a new memo is sent out. Then the sales staff wants to attend, and another memo is issued. Then it's realized that a larger room has to be found, so still another memo. After that, a rainstorm delays everyone's arrival, and so on. An example of memo evolution—

1. Head with mouth and sense organs

2. Backbone

3. Hair and milk glands

6. Large brain and language

5. Upright walking

4. Opposable thumbs and flattened fingernails

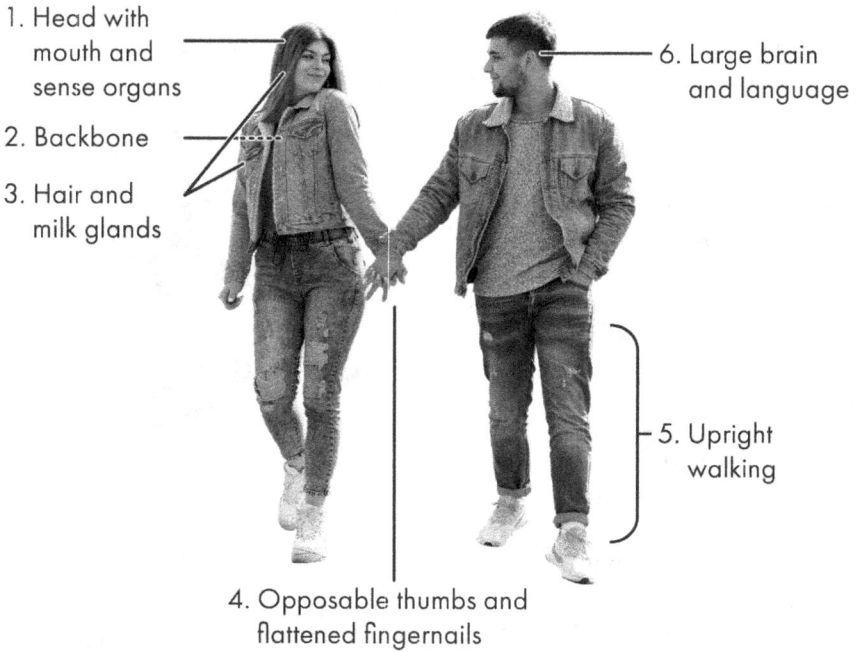

Figure 3. The human body as a patchwork of parts of different evolutionary ages from old to new (1 to 6) (based on a photograph by Gustavo Fring, pexels.com).

Meeting: Room 219, 8:30 a.m.; no. of persons, 10; *Mr. Nelson* presiding

Meeting: Room 219, 8:30 a.m.; no. of persons, *15*; Mr. Nelson presiding

Meeting: Room *105*, 8:30 a.m.; no. of persons, 15; Mr. Nelson presiding

Meeting: Room 105, *9:30* a.m.; no. of persons, 15; Mr. Nelson presiding

Because conditions vary from place to place, different memos apply to different situations. There are earthworm memos, sea urchin memos, mouse memos, human memos, and millions of others. In the long, slow evolution of animal bodies over the course of hundreds of millions of years, environments change. Genetic alterations occur at different times and become permanent in accordance with changed circumstances.

So it is with the human body. On an evolutionary timescale, some parts are very old, some not so old, and others quite new. Our fleshy lips, for example, appeared only relatively recently and are shared by no other species on our branch of the evolutionary tree. Other features have been around for a longer time and are not exclusively ours—fingernails instead of claws or hoofs are a trait we share with monkeys and other primates. Still other features are very old indeed, ones we share with a very wide range of other animals—we share the trait of having hair with most other mammals including giraffes and mice, and we have four limbs, a trait we share not only with giraffes and mice but also with frogs and lizards and birds.

The sequence and timing of when various traits appeared can be determined in either of two ways. One is from fossil bones. Where relevant bones have been found and their ages have been determined by physical and chemical methods, they can tell us when the trait of interest first appeared. For example, the structure of foot bones and pelvic bones, and sometimes even skull bones, can tell us whether an ancient ape walked upright or not.

The other way is by comparing living species. Figure 4 illustrates the reasoning. Even though we talk about modern species having "descended" from ancient species, evolutionary diagrams of this kind usually put ancient species at the bottom and modern species at the top. This topsy-turvy convention may have originated with Charles Darwin's first hand sketch of evolution in 1837, reproduced in Chapter 5, and it persists because we like the tree metaphor.

A new species can arise when a sub-population of an ancestral species branches out on its own; in effect, the ancestral species splits into two. If two closely related species share an inherited trait (such as *c* in Figure 4), one that's absent from other species, then that trait must have arisen on a fairly recent part of the evolutionary tree. If a trait is shared by a wide range of related species (such as *b* in Figure 4), then that trait must have first appeared on an earlier branch. If a very wide range of related species share a trait (such as *a*), it must have originated even further back. And if a trait (such as *d*) is unique to the species, it must have arisen quite recently, in evolutionary terms. The diagram shows how, from the distribution of traits among closely and more distantly related living species, we can conclude that the human traits arose in the order *a-b-c-d* in our evolutionary past.

(Of course, other species on their own branches of the evolutionary tree will have evolved their own sets of traits too, different ones, many of which will not be shared with humans at all, such as laying eggs or having wings. Those species will have to make up their own evolutionary trees for their own traits.)

Such patterns of trait-sharing give only relative evolutionary times, not absolute times. Real times can be obtained indirectly from knowledge about when a particular animal group arose during the course of evolution, determined either from dated fossils or from the ages of rocks where the fossils were found or alternatively from estimates based on known rates of genetic change. That is how, for example, we guess that the ability to

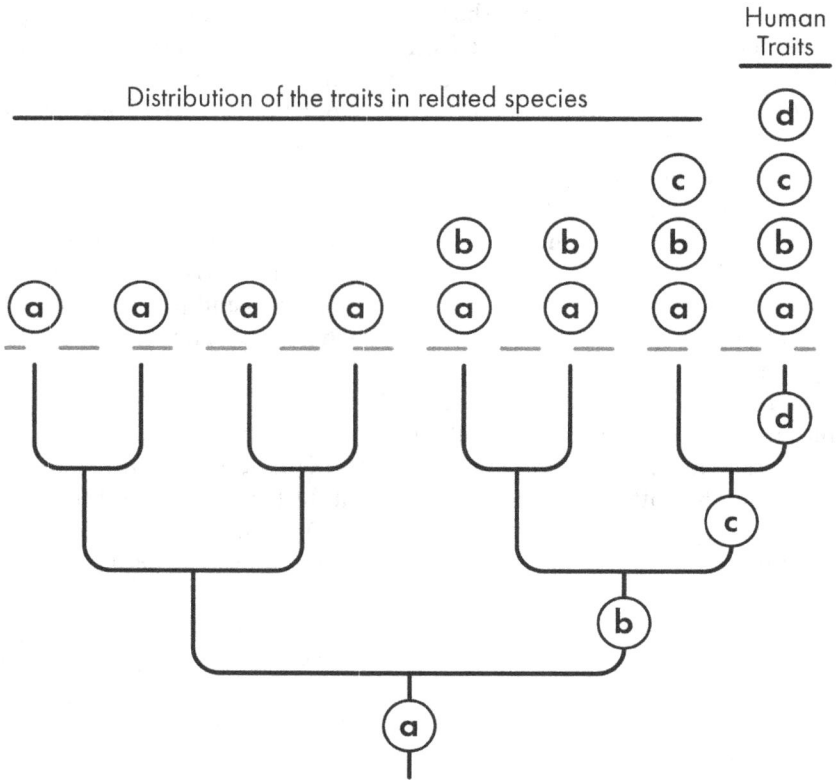

Figure 4. How an evolutionary tree can be constructed from the distribution of traits in living species. Above the dashed line: some human traits, a-d (at right), and the distribution of those traits in other species, some closely related (adjacent to humans) and others more distantly related (toward the left). Below the dashed line: an evolutionary tree that explains the distribution, showing when the different traits originated. More ancient human traits that arose early in evolution (a, b) are more widely distributed, while more modern human traits that arose later (c, d) are restricted to humans and close relatives.

maintain a high body temperature first arose around 250 million years ago, because that trait is common to all living mammals, and from the fossil record that's when mammals first appeared.

This reasoning is also why we surmise that bilateral symmetry—the right side of the body is the approximate mirror image of the left side—is one of humankind's oldest anatomical features. More than 99 percent of all the animals on earth, from the lowliest flatworm to the Maine lobster, from the regal bald eagle to the African elephant, have bilateral symmetry. It's a feature we also share with the fossils of ancient trilobite arthropods from half a billion years ago. It must have arisen at least that long ago and been passed on from some very distant ancestors to most of the later descendants.

The following table shows the approximate ages of some of our different features.

Feature	Approximate time of origin (millions of years before the present time)
Ribosomal RNA genes*	3500
Sec23/24 genes**	2000
Bilateral symmetry	550
Hox genes***	500
Backbone	500
Four limbs	400
High body temperature	250
Placenta	150
Hair	150
Milk for young	100
Fingernails	80
Human-style color vision	45
Upright walking and running	4
Prominent chin	½
Human form of *FOXP2* gene	½

* a class of genes used in the basic cell function of protein synthesis
** genes widely used in internal-membrane function
*** genes used in the early stages of embryonic development

In the same way, the old and the new can be seen in our genes. In their DNA sequences, genes have an "anatomy" too. Surveys of different living species reveal how similar or different their genes are from ours. The human version of a particular gene is more similar to that of a chimpanzee's and less similar to that of the corresponding gene in a bat or a rabbit. Therefore, the relative evolutionary ages of our genes can be determined, and evolutionary trees can be constructed for genes, in the same way as are the evolutionary trees for other parts of our anatomy. An example is given in Chapter 5, Figure 21.

Genetic and anatomical comparisons of living species, combined with fossil evidence, give us a broad evolutionary scheme for all the animals with backbones, the vertebrates (Figure 5).

Often during the course of evolution old parts are modified and put to different uses. When some ancient reptile-like animals evolved into early mammals, there was a shift in the bones connecting the lower jaw to the skull: two of the bones were modified for hearing and became two of

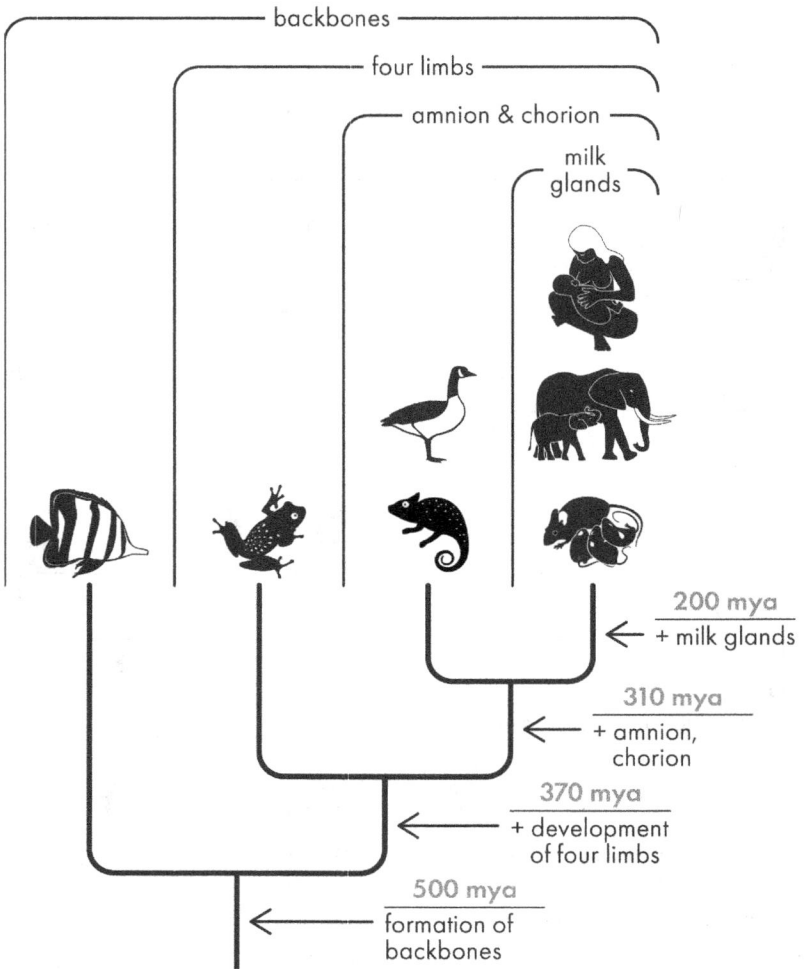

Figure 5. An evolutionary tree of the vertebrates, showing the approximate time of origin of different groups' defining traits (fish, amphibians, reptiles and birds together, and mammals). The amnion and chorion are embryonic membranes which only reptiles, birds, and mammals have. Note that each evolutionary innovation is an add-on, not a replacement, for what went before (mya, millions of years ago).

the three small bones in the mammalian inner ear. The forelimbs of early land animals were half of their four-legged anatomy, but once our ape-like human ancestors began walking upright the forelimbs were gradually modified and came to be used for making tools, carrying food, and throwing stones and spears.

The same principle—old features being cobbled together with new—applies also to our psychology and our behavior. Self-preservation behavior, as when I instinctively jump out of the way when a car runs a red light, must be very old. All animals exhibit it: sponges close up when touched, earthworms contract, clams clam up, cockroaches scurry away into dark corners. Such behavior must have been built into the primitive nervous systems of the very earliest animals. It makes obvious evolutionary sense.

Sleeping at night, vocal expression of pain, and thievery are probably more recently evolved. Such behaviors are found mostly in birds and mammals, which evolved relatively late in the history of life, maybe about 100 million years ago. Also common among mammals, especially in the more social species, are territoriality, dominance hierarchies, personal recognition, learning by imitation, cooperation, maternal instincts, and play activities among the young. These are more recently evolved behaviors.

The primates are a group of animals that first appeared on Earth about 70 million years ago. We human beings, only about 300,000 years old, are latecomers to the primate party. Primates have their characteristic patterns of behavior. They include defense of property (seen in monkeys, baboons, and many humans), practical joking (chimpanzees do it), and tool-making—an ability that is seen in simple forms in monkeys and apes and which is highly developed in humans, especially as a cooperative activity. The instinctive and unreasoning taking of revenge is also typical primate behavior. Monkeys of several different species often hold a grudge and take revenge against others of their species who had previously attacked them. Vervet monkeys, following a Hatfield–McCoy plan of action, may threaten and attack a relative of an original aggressor. A chimpanzee attacked by several other troupe members may later corner one of them when he's alone and give him a thrashing.

We humans can empathize. Observing the behavior of monkeys, we might say, "Look, monkeys behaving like people! How like us they are!" Adopting a more evolutionary view we would say, "Look, monkeys behaving like people! How monkey-like we are!"

Some behaviors and some patterns of thinking are observed only in the recent human line, including some of our now-extinct evolutionary cousins the Neanderthals. There is archeological evidence that Neanderthals had cooking, primitive hygiene, representative art, religious rituals, and perhaps music and dancing too. Characteristic of modern humans only are counting, time measurement, moral codes, complex language, a concept of fairness, and a recognition that death will come to us all (although elephants and chimpanzees might have this awareness too). We might be the only species on earth that can suffer from boredom. These are behavioral traits more modern than revenge-taking or tool-making. Of course, much of this can be attributed to cultural advance and not gene evolution, but underlying it all is increased cognition, based on the evolution of the brain.

In all this, our behavior, like our anatomy, consists of a patchwork of old, newer, and newest.

1

The Human Machine

...if any sensations enter in violently from without and drag after them the whole vessel of the soul, then the courses of the soul, though they seem to conquer, are really conquered.

—Plato, *Dialogues* (4th century BCE)

The 17th-century American poet Anne Bradstreet described the dual nature of the human being in "The Flesh and the Spirit" in this way:

> I heard two sisters reason on
> Things that are past and things to come.
> One Flesh was called, who had her eye
> On worldly wealth and vanity;
> The other Spirit, who did rear
> Her thoughts unto a higher sphere.

"Sister," quoth Flesh, "what liv'st thou on
Nothing but Meditation?
Doth Contemplation feed thee so
Regardlessly to let earth go?
Can Speculation satisfy
Notion without Reality?
Earth hath more silver, pearls, and gold
Than eyes can see or hands can hold.
Affects thou pleasure? Take thy fill.
Earth hath enough of what you will.
Then let not go what thou maist find
For things unknown only in mind."

Spirit—"Be still, thou unregenerate part,
Disturb no more my settled heart,
For I have vowed (and so will do)
Thee as a foe still to pursue,
And combat with thee will and must
Until I see thee laid in th' dust.
Sister we are, yea twins we be,
Yet deadly feud 'twixt thee and me,
Mine eye doth pierce the heav'ns and see
What is Invisible to thee.
If I of Heav'n may have my fill,
Take thou the world, and all that will."

Is there really a spirit (or mind, or soul) that is distinct from the flesh? Does something have to be added to the physical and chemical constituents of the human body to make it fully human? Or are the physical and chemical components by themselves enough?

What Do People Think?

Not long ago I asked 99 people, ranging in age from 18 to 87, some with technical backgrounds and some without, how they would answer this question:

> If a body in the shape of a human being were assembled by artificial means carefully enough, cell by cell and molecule by molecule, in such a way as to give it all the right molecular structure, brain-cell connections, and appropriate physiological responses, so that there were no physical tests, no answers it might give to your questions, that would reveal its origin—would it be fully human?

Forty-two people answered *yes*, and 57 answered *no*. There was a tendency for scientifically trained people to say *yes* and a tendency for older people to say *no*. Those answering yes gave these explanations:

- Everything that is human is based on our molecular structure.
- The manner of functioning is what makes us human, not how we were created.
- If it walks like a duck and quacks like a duck, then it's a duck.

Among the reasons given by those who answered *no* were the following:

- How a structure is created is an essential part of what it is; to be fully human, it would have to go through the normal processes of embryonic development, spend nine months in a womb, and be cared for by a mother after it came into the world.
- The human body is just the vessel for the soul; you can create the body however you like, but if it doesn't have a soul, it isn't human.
- It would be like Cheez Whiz, artificial, not human.

But truth is not determined by opinion polls. If it were, then at one time the Earth really was flat and really was located at the center of the Universe.

People versus Machines

Some functions that at first glance seem uniquely human can also be carried out by machines made of metal and plastic. A machine has been constructed that can recognize itself in a mirror. It uses the camera eye to record a video of itself in the mirror and then compares the video record with its internal diagram of its own structure and the program of its own movements, such as waving an arm. If they match, the machine registers "self"; if they don't, it registers "other." With a human being there are a greater number of matching clues in the image and a greater number of recordings in memory, but is a woman checking her mascara in a mirror, or a man using his reflection to straighten his tie, really doing anything different?

Young children learn to write by using their brains to coordinate what they see with their eyes and how their muscles are moving their hands. Similarly, a machine provided with a camera, a mechanical arm, and the appropriate computer program to connect the two can learn how to reproduce faithfully what are initially unfamiliar written symbols. It can learn from experience.

A high-tech toy with wheels, several small motors, some miniature sonars, internal and external sensors, and a small computer to integrate all the operations can learn to navigate around a room without bumping into the furniture. Initially it runs into things. The shock and the location are recorded internally, and the information is relayed from the sonars to the computer, so that afterwards new electrical signals are sent from the computer to the motors, which steer the toy in a different direction. The toy can quickly accommodate its behavior to the presence of a new table or chair. Therefore, not every aspect of its behavior need be programmed in advance; it can respond to changing circumstances.

Humans can do such things with exquisite delicacy and control, which robots can only crudely imitate. But humans have had the advantage of hundreds of millions of years of the tinkering and fine tuning that comes with evolution.

It is not only the long span of evolution that is on our side. The subtlety and nuance of human behavior are also a function of the materials we're made of: organic molecules dissolved in lots of water (about 70 percent of our body mass) with a sprinkling of inorganic salts. At the molecular and cellular levels, we are made of hundreds of thousands of different parts, which are continually and automatically being replaced as they wear out and as the body adjusts to different circumstances. The consequence is that we have flexibility, adjustability, and responsiveness to the situation that machines can only dream of. We are soft, wet, and malleable. Machines, so far, even sophisticated ones, are hard, dry, and rigid. They have relatively few parts and permanent ones upon which they can depend only so long as the parts don't wear out. But it's not impossible to imagine a soft, self-renewing, capable robot of the future which (who) might come close to imitating much of what humans can do.

Machines are assembled consciously and deliberately piece by piece, either by humans or by machines made by humans. But in our offspring, we humans produce approximate copies of ourselves without conscious planning of the construction procedures, blindly following genetic instructions honed over countless years of evolution, on tracks laid down by millions of prior generations. Even as individual human beings reach maturity and begin their inevitable march toward death, new human embryos are gradually unfolding inside their wombs. This may be the one way in which humans might forever distinguish themselves from machines.

The human body is a machine in another fundamental way, the way a refrigerator or a printing press or an automobile is a machine. Everything it does is the result of the interactions of its components with one another and with the environment. What we view as special attributes distinguishing a human from a machine are simply the consequences of our being more complicated, by orders of magnitude, than other machines we're familiar with. If machines could be made to be sufficiently complex, and flexibly programmed in the right way, might they not also have all the traits we think of as exclusively human? We might be just that kind of machine, not a machine to which something else has been added to make us human. The issue has been debated throughout recorded history.

Mind, Spirit, Soul

London's influential *Quarterly Review*, commenting on Mary Shelley's 1818 novel *Frankenstein*, expressed its opinion in an 1820 editorial as follows:

> The most probable conclusions to which our reason can carry us [is] that life, in general, is some principle of activity added by the will of Omnipotence to organized structure,—and that, in man, who is endowed with an intelligent faculty in addition to this vital principle possessed by other organized beings, to life and structure an immaterial soul is superadded.

There are some ideas that act forcefully on my mind to make it easy to suppose that something has to be "superadded" to mere matter to make it human, that the spirit—or soul or mind—is something very different from the physical body that houses it. The 17th-century French philosopher and mathematician René Descartes famously proclaimed a dualistic view by saying, "I have a distinct idea of myself as a thinking thing occupying no space; on the other hand, I also have a distinct idea of my body as an unthinking thing occupying space; therefore it is certain that my mind, by which I am what I am, is entirely and truly distinct from my body."

Take the spirit of creativity. A musician, a painter, or a poet all express their creative spirit from somewhere inside, which has nothing to do with the outward appearance of their body. If you'd ever seen or heard Alice Sara Ott play *Grieg's Piano Concerto in A minor*, you'd have witnessed human spirituality on its highest plane. You don't think of the piano keys and the strings. Or when you read Elizabeth Barrett Browning's "Beholding, besides love, the end of love, / Hearing oblivion beyond memory; / As one who sits and gazes from above, / Over the rivers to the bitter sea," you don't think of the nerve impulses from the brain or the muscles and tendons that move the pen over the paper. They are something different. You could list all the vowels and consonants, all the sound frequencies from the piano strings in cycles per second; you could read all the results of magnetic resonance imaging studies of the brain and talk about the interconnections among billions of brain cells—and it wouldn't be poetry and it wouldn't be music. Human creativity must come from somewhere else, something you can only call the human *spirit*.

A beautiful body can hide an ugly soul, and an ugly body can hide a beautiful soul. There is no correlation between the two. As Christine de Pizan wrote, "You very often see men who are well built and strong yet pathetic and cowardly, but others who are small and physically weak yet brave and tough. This applies equally to other moral qualities." The spirit and the body are not the same. And so with any one individual at different times. My spirit can fluctuate wildly, joyful one minute and gloomy the next; my mind can be focused one minute and a moment later be drifting off into gauzy daydreams—all the while my body remains the same.

I have a subjective mental life that it seems no machine could have. It's not equivalent to any tracings of my brain's electrical impulses. Last week I caught a passing whiff of perfume and was reminded of an old girlfriend, a rose garden, a beach on an island. The memories themselves may have a physical basis, but the subjective experience of having the memory was purely mental. A robot's computer-brain might have parallel chemical and electrical events implanted and wired into it, and the robot, zombie-like, might even report them, but the robot wouldn't *have* the experience of the memory.

There's also free will. I can make choices of my own volition. I don't have to be a slave to biological instincts. I can choose pain and keep my appointment with the dentist. My biology may tell me that life is all about having children, but on any given evening I can still decide to use birth control. If my body craves an extra piece of pie, I can still forego it. Of course, in all these cases there are competing desires: pain in the dentist's chair tomorrow vs. worse pain next month if I don't go and so on. But if I were a machine my behavior would be completely determined by how I was constructed in the womb, the commands that were programmed into me, and how rationally I processed new information coming in from my environment. But those things don't describe how I, as a human being, run my life. The freedom is in the internal events in my mind, perhaps unknowable, and in their independence from external constraints.

Then there's the biggest change of all, the transition from life to death. The physical body has the same structure just after death as just before; something non-physical appears to have happened right at the moment of dying. It seems as if at that instant something has departed, leaving the body behind. This is a belief that humankind probably held even before there were recorded depictions of it, as in the 3000-year-old Egyptian *Book of the Dead*, showing the soul as a bird-like creature hovering over the dead body (Figure 6, upper panel).

What do we really know about these things?

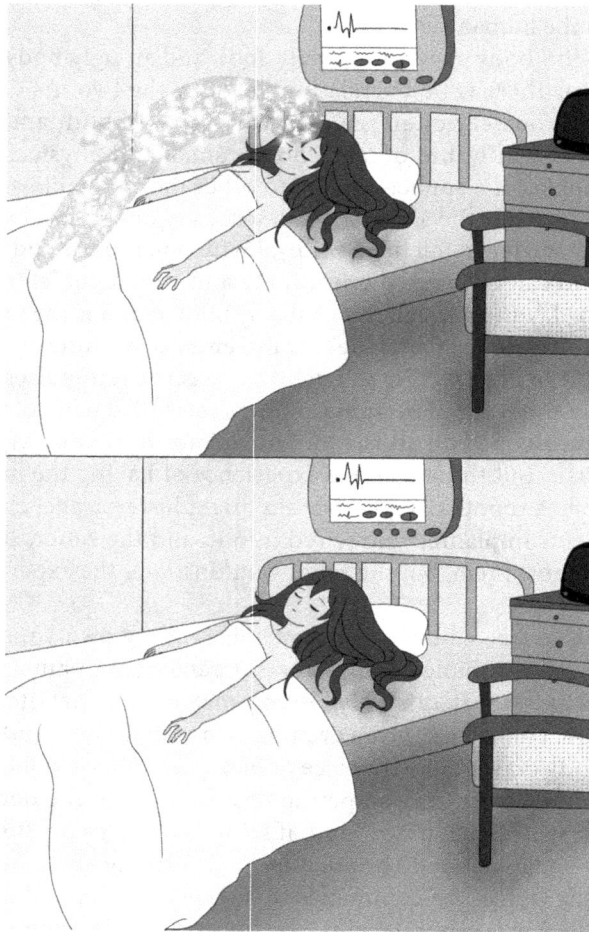

Figure 6. Two views of the moment of death. Upper panel, the soul is leaving the body. Lower panel, the machine that is the human body simply stops working, as the monitor shows (illustration by Makinze Jackson).

Matter

The more we learn about human attributes such as love, curiosity, intelligence, creativity, and self-awareness—and the more we learn about the details of brain function—the more probable it seems that all these human traits could probably be programmed into a sufficiently sophisticated machine. Love is internal chemistry and nerve-cell activity. Curiosity is behavior directed by genes and proteins and neurotransmitters. Intelligence is a gradually evolved brain function, individual aspects of which can be mimicked by electronic devices. Self-awareness is a general awareness with a mirror-like feedback function. And so on. Those who deny the human being's machinelike nature can take recourse in maintaining that the human being has a soul, while a machine however complex does not. But it may be that the *idea* of a soul or spirit is itself only a convenient myth, one which we use as a kind of shorthand to help us account for all the more complicated features of human consciousness and behavior. What we call our "spiritual side" may just arise naturally from all the complex interactions of the molecules, genes, cells, tissues, and organs of which we are composed. From this comes the evolutionary story of this book.

Similarly unpredictable is the weather. We now have, as a basis for comprehension, general principles based on the physics and chemistry of the atmosphere, the configuration of the terrain, the properties of the oceans, pressure gradients, temperature variations, and computer models that mimic the interactions of all the variables. This knowledge is more consonant with the rest of our knowledge about how the world works. We no longer invoke angry deities and rain gods and bolts of lightning hurled from heaven to provide our explanations.

When the heart fails and the blood stops circulating, the delivery of oxygen to the body's cells and tissues ceases. Energy transformation in the cells fades away almost immediately, like a candle that sputters and goes out when the wax is used up. Within seconds the delicate cellular membranes and transient protein structures, previously maintained by a continuous flow of energy, begin to disintegrate. The fragile molecular organization sustaining the body's operations begins to unravel. Human life as we saw it a few moments before—breathing, responsiveness, eye movements, facial expressions, speech—ceases (Figure 6, lower panel).

As soon as the machinery of the body stops working, the large molecules on which it depends begin to disintegrate. Some of the molecular fragments will disperse into the atmosphere, or dissolve in lakes and rivers and rain, or be compacted in soil fragments; some will be recycled, taken up and used by bacteria and worms and fungi and plants for their own purposes. But that particular human life will be over.

The idea of *the soul* is a useful abstract concept covering different aspects of human experience, but it's not a real object that enters the embryo early on and then leaves the body at death; it's something created by the human mind. When I turn off the ignition of my car I don't think, "The car's motion has departed."

It's true that a person's personality is not usually revealed by their physical appearance; a person beautiful on the outside is not necessarily beautiful on the inside. Nor are one's fleeting thoughts and feelings and dreams accompanied by obvious changes in physical appearance. But we're looking in the wrong place. If we want to assess the state of someone's digestion, for example, we don't look at the color of their eyes and say, "Oh, there can't be a physical basis for digestion." To see the real physical basis for someone's personality, or their passing thoughts and their changes of mood, we'd have to see inside. Maybe we could do that if people walked around with a special kind of screen in their foreheads; then we could see the physical events inside the brain, looking like the flickers of lightning in storm clouds, and perhaps learn how to read them. Those are the mental events we refer to as "non-physical." But thoughts and moods are physical events, and physical and chemical states, and they have a physical basis.

In the 5th century BCE, Empedocles wrote, "Nourished in a sea of churning blood, where what men call thought is especially found—for the blood about the heart is thought for men." He may have got the anatomy wrong, but he had the right physical idea.

My subjective life, my inner mental life, is so personal and private and so unlike anything I can imagine a machine experiencing that it seems to set me off from any purely mechanical device. But I can't be sure. What seems to distinguish people from machines is primarily the uniqueness of each person in contrast to the sameness of each machine (of a given type). We know the main sources of individuality in people. It is each person's unique combination of genes and the proteins the genes code for, and the unique combination of experiences they had while growing up. But machines are deliberately made to be all the same; to be composed, as much as possible, of identical parts; and to perform, as much as possible, in exactly the same way regardless of past operations. If each individual machine were deliberately constructed to be different from every other machine of that kind, and if each such machine had an internal record of its own past experiences and performances, then those machines would be more like people. In the distant future, some robot's 86 billion microchips crammed into its braincase might have built into them unique fictional mental associations, fabricated feelings, and false memories of experiences it never had, all interconnected with one another, with most of them capable of being recorded and perhaps expressed through a playback device operating at millisecond timescales.

Isn't it in just those interconnections and the playback to consciousness that my unique subjective experiences reside? The robot's private mental life, to the extent it would be aware of it—and why not?—would be indistinguishable from that experienced by any normal human being. Perhaps I am that robot.

As for free will—it may be an illusion. It is not a question of merely being free from external restraints. As Sigmund Freud pointed out, our unconscious desires and fears determine our choices much more than we're aware of. We're not cognizant of the details of what's going on behind the scenes, details that may be determining our behavior.

More importantly, what goes into my day-to-day decisions could be a combination of the physiological responses and instincts I inherited as a member of the species *Homo sapiens*, my genetically controlled fetal development, the socialization I experienced as I was growing up, and how I am evaluating my present circumstances and judging my future prospects. All these factors, down to their tiniest details, could go into determining my day-to-day decisions. I could just be a very complicated, hard-to-predict machine. Note that *unpredictable* is not the same thing as *free*. Pull the handle of an old-fashioned slot-machine—the final position of any of the three reels cannot be predicted because of all the minute details that determine exactly where each stopping device engages a notch in the fast-spinning reel. The system is mechanically determined, even if unpredictable. My so-called "free choices" could be determined in the same way. Minor details of brain function, subtle evaluations of risks and rewards based on early training, past memories, present circumstances and predictions for the future all together determine my choices. But they're still determined, and there is no real freedom in the sense that most people conceive it.

I may not be much different from my dog. His intentions are determined only by his experiences, the current circumstances, and the way his dog-brain was wired from the beginning. The only difference is that I like to think I'm making choices, while Fido doesn't care about that question one way or another.

The assertion of a purely physical origin of "spirit" does not *refute* the idea of a spirit separate and distinct from the physical human body—same observations, different interpretations. But the so-called "evidence" for a separate spirit is not evidence for anything at all, because a purely physical/chemical interpretation is available to explain the same observations. There is so far no real evidence for an independent spirit in the human frame, no evidence for a "ghost in the machine," to use Gilbert Ryle's phrase. The physical interpretation is consonant with everything we know so far about life in general—that it can be broken down into its chemical parts in test tubes and Petri dishes and that under those conditions the parts do exhibit their separate functions. The more complex features of human brain function are amenable to study, and they also show promise of being explainable, at least in principle, by purely physical and chemical processes. It may be that the human spirit, so apparently inexplicable, arises from no other source than the complexity of the interactions of the body's component parts.

How can lifeless molecules be assembled to make a living human being, without there being something extra added? Imagine a bunch of molecules with their individual shapes and surfaces and chemical properties, like lifeless pieces of a jigsaw puzzle (Figure 7). Now join them together in the right way, and something new emerges. It is the assemblage that counts, that creates something new, even though nothing new has been added. In human life, as in all other life, there does not even have to be an outside Force or Presence that puts the pieces together in the right way; being of the right shapes, they assemble themselves.

With this provisional conclusion in hand, we can see the heightened significance of genes and proteins and other molecules: fundamentally, that's all there is. How they all interact, and how they all add up to human life and the human experience, remains the hard problem.

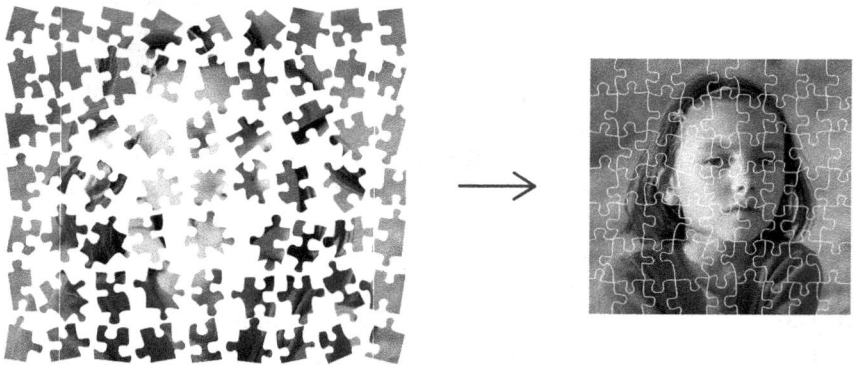

Figure 7. This figure illustrates how, when the parts in the left and center panels are combined, something apparently new emerges (right panel), even though no new ingredients have been added (photograph of the author's granddaughter, Mariel Montero, by Richard Baron).

2

Cells, Genes,
and Other Small Parts

*I told several lines of these pores, and found that there
were usually about threescore of these small Cells placed
end-ways in the eighteenth part of an inch in length,
whence I concluded there must be neer eleven hundred of
them, or somewhat more than a thousand in the length of
an Inch, and therefore in a square Inch above a Million,
or 1166400. and in a Cubick Inch, above twelve hundred
Millions, or 1259712000, a thing almost incredible, did not
our Microscope assure us of it by ocular demonstration.*
—Robert Hooke, *Micrographia: or Some Physiological*
Descriptions of Minute Bodies Made by
Magnifying Glasses—with Observations
and Inquiries thereupon (1665)

Contrary to what the early 19th-century nursery rhyme says, it's not
snips and snails and puppy dogs' tails, nor sugar and spice and everything
nice, that the human body is made of. It's cells and macromolecular com-
plexes and intracellular membranes and extracellular matrix and fluids and
DNA and RNA and protein and salts and fats and other small organic molecules. The human body is about 70 percent cells and about 30 percent extracellular materials. The extracellular materials consist of both fluid (blood
plasma in the heart and blood vessels, cerebrospinal fluid in the brain and
spinal cord, and watery solutions in the narrow spaces surrounding cells and
a few other places) and solid (the extracellular matrix of bones and cartilage,
and fibers and gels filling the small spaces throughout the body).

Cells

Being made of cells with their internal organization is one of the oldest
features of human beings. We share that basic organization with all other

organisms, although bacterial cells are built on a slightly simpler plan. The structure is that of the "eukaryotic" cell (Greek, *eu-*, "true" + *karyon*, "nucleus")—that is, a cell with a nucleus, visible even with a microscope of only moderate magnification. These days, with several decades of electron microscopy behind us, we know much more about the internal organization of cells. Figure 8 illustrates some of the internal structure that can be seen with the electron microscope. Some of the internal elements (labeled) and their descriptions below give an idea of the things our cells can do.

- The **plasma membrane** regulates the movement of small molecules into and out of the cell and relays molecular signals from the outside of the cell to the inside.
- The **Golgi apparatus**, named for the Italian physician and microscopist Camillo Golgi who first observed it in 1898, is an internal membrane system mostly involved in segregating and packaging of protein molecules to be secreted by the cell to the outside.
- **Ribosomes** are complicated little structures where proteins are synthesized; that is, where specific gene messages from the nucleus are translated into the amino acid sequences of proteins. They occur by the thousands and even millions in some cells. Some ribosomes are attached to internal membranes, and the proteins

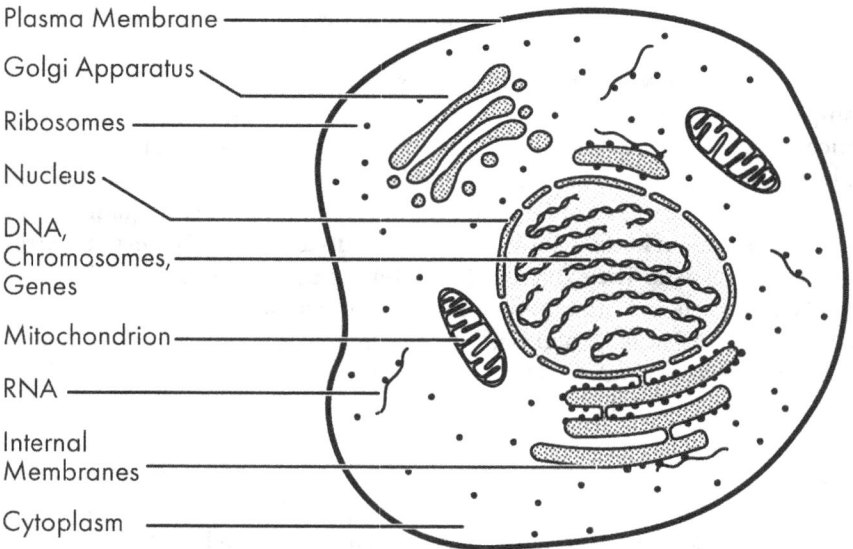

Figure 8. Diagram of a generalized human cell. See text for a description of the different parts.

they make end up in the plasma membrane or are secreted to the outside of the cell. Some ribosomes are free in the cytoplasm (the region inside the plasma membrane but outside the nucleus); the proteins they make are used internally by the cell.

- The **nucleus** is where almost all the DNA (the molecule that makes up genes and chromosomes) is located; it is bounded by a membranous nuclear envelope.
- **DNA** (deoxyribonucleic acid) is the molecule that **genes** are made of.
- **Chromosomes** are long strings of genes.
- **Mitochondrion** (pl., mitochondria) is where important energy transformations occur, producing high-energy molecules that are used everywhere in the cell. The mitochondria contain a few genes of their own.
- **RNA** (ribonucleic acid) has many functions, including the important one of being gene copies that a cell makes in the nucleus and then sends out into the cytoplasm, where in association with ribosomes they are used to make proteins.
- **Internal membranes** include the extensive membrane system known as the endoplasmic reticulum, which segregates different proteins destined for the plasma membrane and for secretion to the outside.
- The **cytoplasm** is a general term for everything inside the cell but outside the nucleus.
- **Proteins** comprise almost all the working molecular machinery of a cell. Every cell has thousands of different kinds of protein molecule and many millions of individual protein molecules. They fill up the whole cell, making up half of the cell's non-aqueous mass.

Every cell has the same genes, but during the formation of the embryo and the fetus and finally the adult, different cells in the body come to use their genes differently, changing their protein content and many details of their internal organization. Some kinds of cell are actively and continuously synthesizing proteins and so have huge numbers of ribosomes, while other cells are quieter and have fewer ribosomes. Some cells do a lot of secreting and have a lot of internal membranes (endoplasmic reticulum and Golgi apparatus), while other cells keep most of their proteins inside and have fewer internal membranes. It's the proteins that make one cell function differently from another—nerve cells, muscle cells, liver cells, kidney cells, skin cells, blood cells, bone cells, and hundreds of other kinds (see Appendix I).

Aside from genes' being *used* differently in different cells of the body, the variety of information in genes *in different people* generates variety in their proteins. Protein variation from person to person makes their cells behave differently, and how cells behave (in the aggregate) determines the details of how the body's tissues and organs function, and that in turn determines how the body as a whole is structured and how it functions in dealing with the outside world. And *that* influences that particular body's chances of surviving and passing on its genes to its offspring. The chain of causation from gene to survival is long and complicated.

The eukaryotic cell is a near-miracle in its complexity—not only in its internal organization, but in the way it creates embryos and then adults with their hundreds of interacting cell types. The cell's internal compartmentalization allows it to allocate different kinds of protein to different places in the cell. Some proteins are retained in the cytoplasm for specialized function or for an internal structure that allows the cell to change shape or migrate from one place to another in the embryo. Regulatory proteins are transferred from the cytoplasm back to the nucleus; they determine how and when particular genes are used. Some proteins are sent to the plasma membrane, where they mediate interactions with adjacent cells or with the non-cellular matrix in which they are embedded. Some proteins are exported as secretions that benefit the body as a whole.

The earliest cells, once having achieved the eukaryotic-cell kind of internal organization, were served so well by it that it was passed on to all the life forms appearing later: mice, whales, flies, oak trees, dandelions, mushrooms—all the plants and animals and fungi on the planet, including, at a late date, us humans. The basic eukaryotic organization of our cells must therefore be very ancient—by some guesses at least two billion years old, half the age of the Earth itself.

Molecules Small and Large

All the cells in the adult human body come from the repeated divisions of the original single cell, the fertilized egg: one cell to two, two to four, four to eight, eight to 16, and so on, ending up after about two decades with approximately 10 trillion cells in all. Water, H_2O, is the main molecule of the typical cell, making up about two-thirds of its mass. We land animals are overly conscious of our solidity and the relative dryness of our external surface, but inside we are like sopping wet sponges. Our watery internal environment is a carry-over from our distant origins, the first primitive cells that billions of years ago floated in primeval lakes and oceans. We still carry those ancient seas within us. Besides being a medium in which other molecules are dissolved, water helps to maintain the molecular structure of larger molecules such as proteins and DNA. Water promotes interactions between molecules and stabilizes cell membranes. It's not just a passive bystander either; it participates directly in many of the cell's chemical reactions. It's hard to imagine any kind of life without water. Water is the first thing we look for in the search for extraterrestrial life. There may be water on planets circling other stars, and some of the moons of Jupiter and Saturn have water, so perhaps there's life of some kind in those places.

The one-third of a typical cell's mass that is not water consists of lipids, sugars, amino acids, organic acids, other small molecules, salts … and two kinds of very large molecule, the nucleic acids and the proteins.

The nucleic acids—deoxyribonucleic acid (DNA) and the chemically related ribonucleic acid (RNA)—are informational macromolecules. Genes are made of DNA. By and large, DNA doesn't *do* anything; its use is in storing and copying information. The information in the DNA is read by the cell, just as the information in a recipe is read by the chef (Figure 9). One difference is that the recipe is stored *outside* the chef, while the DNA is stored *inside* the cell.

Proteins—long, coiled-up, folded strings of amino acids—play a different role in life. Proteins are the functional macromolecules of the cell; they actually *do* something. They constitute most of a cell's basic structure, and they perform most of the work in carrying out the functions of the tissues and organs of the body. There are tens of thousands of different proteins, each with its own specific chemical structure and role to play. Structural proteins provide the structural framework of cells; enzyme proteins direct the chemical changes of various small molecules in the body's metabolic pathways; transport proteins serve as the molecular gateways and channels regulating the passage of small molecules and ions across cell membranes; motor proteins are responsible for the movements of cells; regulatory proteins are part of the molecular machinery that determines which particular genes are used, and when, in a particular cell type.

Chromosomes, Genes, and Proteins

Your DNA, which is what genes are made of, is good for storing information and making copies of itself, but it's not the kind of molecule that can carry out the biochemical and biophysical processes that support life. It's your proteins that do that. You might get along for a while without your DNA (as your circulating red blood cells do, for their four-month lifetime), but not without your proteins. Which would you rather eat, the recipe or the stew? Although you might not be able to make a stew without the original recipe (the genes), it's the components of the stew itself (the proteins) that sustain life.

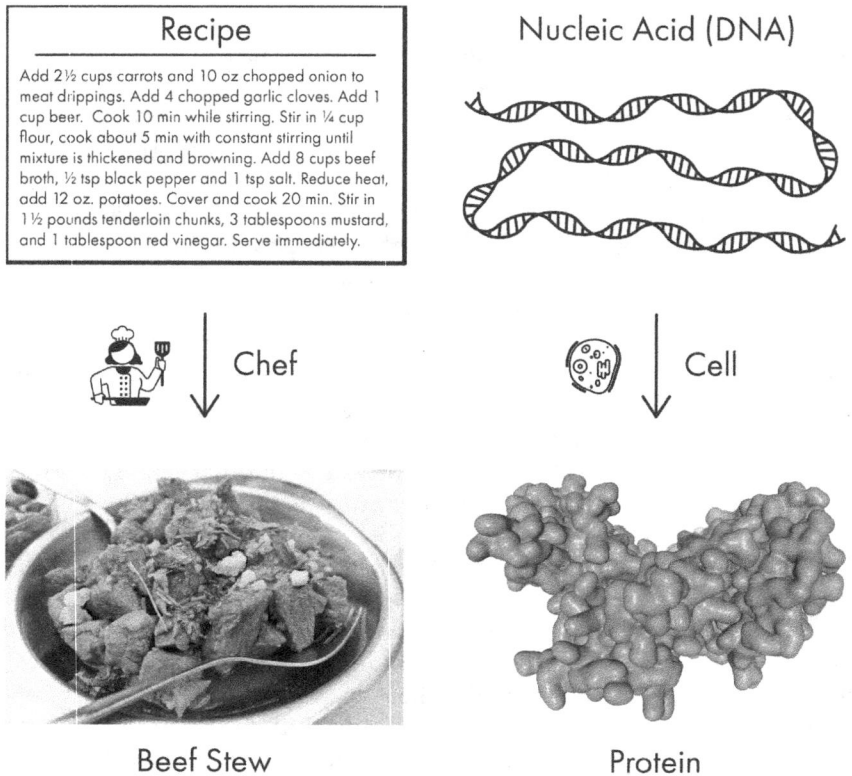

Recipe
Add 2½ cups carrots and 10 oz chopped onion to meat drippings. Add 4 chopped garlic cloves. Add 1 cup beer. Cook 10 min while stirring. Stir in ¼ cup flour, cook about 5 min with constant stirring until mixture is thickened and browning. Add 8 cups beef broth, ½ tsp black pepper and 1 tsp salt. Reduce heat, add 12 oz. potatoes. Cover and cook 20 min. Stir in 1½ pounds tenderloin chunks, 3 tablespoons mustard, and 1 tablespoon red vinegar. Serve immediately.

Nucleic Acid (DNA)

Chef

Cell

Beef Stew

Protein

Figure 9. The gene (made of the nucleic acid, DNA) is a specific recipe used to make a specific kind of protein, the way another kind of recipe is used to make a specific food dish. The DNA in a cell is read by the molecular machinery of the cell itself, while the beef-stew recipe is read by a chef. A difference is that the food recipe is stored *outside* the chef, while the protein recipe (the DNA) is stored *inside* the cell (image of protein molecule based on D. Sehnal, A.S. Rose, J. Kovca, S.K. Burley, S. Velankar, 2018, Mol*: Towards a common library and tools for web molecular graphics, *MolVA/EuroVis Proceedings*; doi:10.2312/molva.20181103, Protein Data Bank RCSB PDB ID 3BP4).

If each different kind of protein could somehow get itself copied directly from a pre-existing protein molecule of the same kind—perhaps with the aid of some sort of molecular-level 3D copying machine—there might be no need for DNA. But proteins are too complicated for direct copying, their chemistry too dedicated to other functions, and their long chains too bent, twisted, and folded up like molecular origami. Besides, many new proteins appear for the first time during the course of embryological development, when there is no template from which to copy them directly. Even in the adult body, many proteins are unstable and constantly breaking down or being exported from the cells that made them (digestive enzymes, for example), and more have to be re-synthesized all the time from scratch.

The necessary information for making new protein molecules resides in the DNA, among the more important molecular parts of a cell. Genes are not the magical little elements we sometimes imagine, located somewhere in mysterious inner chambers of the body, silently dictating the details of our lives, pulling our strings like puppet masters. Genes are always there in the background, but in the operations of the machinery of the body their effects are realized only indirectly, via the proteins for which they carry the instructions.

Genes are located in the nucleus of a cell (except for a very small number of genes in mitochondria), as segments of very long DNA molecules, the chromosomes. An average-sized human chromosome has about a thousand genes in it. It contains the information for making at least a thousand different kinds of protein (sometimes more if the same gene is used in more than one way). In humans, there are 23 different chromosomes, each different in its gene content—or 24 if you count the Y chromosome, essential for male life but not for female life. In Figure 10 are shown the chromosomes of a cell from a man, as seen under an ordinary microscope on the left, and on the right after the images have been cut out from the photograph and arranged and numbered according to size. (They come in pairs because the man's mother donated one set and his father another set.)

The human number of 23 chromosome pairs is similar to that of many other monkey and apes. Baboons and the Rhesus monkey have 21 chromosome pairs, gibbons 22, marmosets 23. Interestingly, the other great apes, besides ourselves—chimpanzees, bonobos, gorillas, and orangutans—all have 24 chromosome pairs. Two average-sized chromosomes in some great ape ancestor of ours fused end to end to make our second-largest chromosome. Our chromosome no. 2 is therefore a modern structure, even though the genes in it are old, pre-dating the evolution of primates. Our chromosome no. 2 like having a new bureau-drawer for an old pair of jeans.

Among mammals in general, chromosome numbers are more variable: dogs have 39 chromosomes, cats have 19, elephants have 28, and mice have 20. Nevertheless, dogs, cats, elephants, mice, and humans all have about 20,000 genes. In the different species the genes are just distributed among different numbers of packages—different numbers of chromosomes—for evolutionary reasons that are not completely understood.

Each of our 23 different chromosomes has its own content of specific genes lined up in tandem, like the words in this sentence. The whole set of human chromosomes contains an estimated 20,000 genes in all, roughly

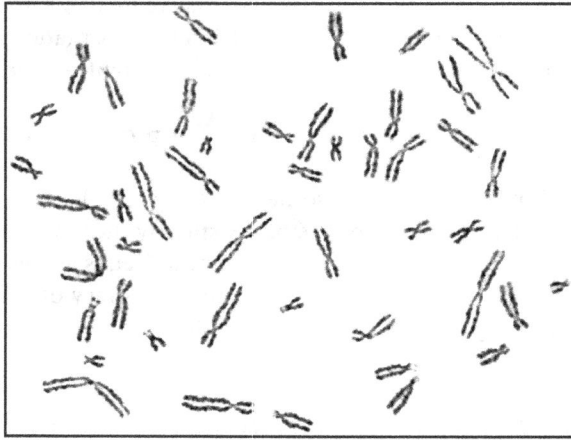

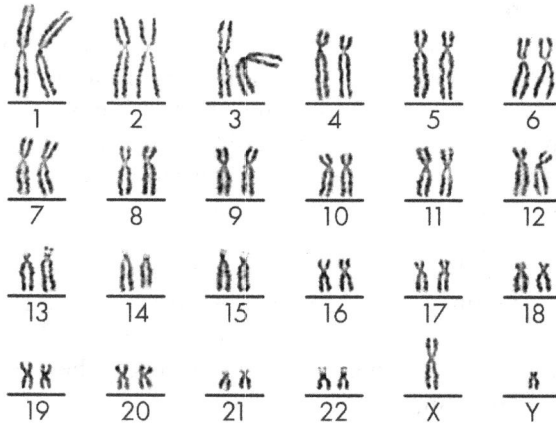

Figure 10. The 46 chromosomes of human cells, numbered according to size. Each chromosome consists of a string of many genes. The chromosomes occur in pairs, one member of each pair donated to the child by the mother, the other donated by the father (courtesy Dr. Yao-Shan Fan, M.D., Ph.D., Department of Pathology and Laboratory Medicine, University of Miami Miller School of Medicine, Miami, Florida).

equal to the number of words in *The Old Man and the Sea*. Any specialized cell, however, uses only a fraction of all the genes it carries, just as a specialized gourmet chef would use only some pages of an all-purpose recipe book.

Once the human genome project had been finished (in 2003), we had a fairly complete inventory of all our genes and their DNA sequences. The list by itself, however, is not very useful. We're still a long way from fully understanding how our cells read their DNA and how their use of their genes leads to a functional human being. For that, we would need to know more details about when the individual genes are used, how all the proteins encoded by the genes interact with one another, and what happens when the whole assembly interacts with the environment. Here's another parts-list that illustrates the difficulty:

☐ Angled lever ☐ Center-rotating handle ☐ Circular dial face ☐ Crank with two projecting pins ☐ Metal housing ☐ Metal penalty flag ☐ Pawl ☐ Pointer ☐ Ratchet wheel ☐ Semi-circular arm ☐ Slotted metal face panel ☐ Slotted slide ☐ Straight lever A ☐ Straight lever B ☐ Timing mechanism ☐ Two-inch straight arm

What are these parts? What do they do? How do they fit together? What do they make when assembled? How does the whole assembly work? What is its function? How does it respond to things coming in from the outside? This happens to be a list of the parts of a standard old-fashioned parking meter. Its behavior depends on its internal structure, on the interaction of its parts, and on the coins that are fed into it. The list of human genes is like that: an essential list, but not enough to predict how the human machine will behave when assembled, nor what it will do when coins are dropped in. Nevertheless, a comprehensive listing of parts is a start, a first earthbound leap toward the stars.

Here are some details about how cells make use of their genes. The details are given in order to take out some of the magic.

A gene is made of a long string of four "DNA bases"—adenine, guanine, cytosine, and thymine, abbreviated A, G, C, T—in a particular sequence that is unique for each gene; for example, a small part of one gene might read

...ATAGCTCCGATAAGCGCGCTGCCCAGTAGGAGCTCCAC...

The information in a gene's DNA specifies the particular protein to be made and also whether or not the gene will be used in that cell at all. In reality a gene is a much longer sequence, thousands of bases in fact. Part of a gene's DNA sequence, the "coding sequence," carries the genetic code, chemical information for how the cell will string amino acids together to make a particular kind of protein. At the beginning of the coding sequence is a regulatory sequence. This, in combination with certain regulatory proteins, determines if and when the cell in question will in fact be using that gene, and if so, how much protein will be made. Figure 11 below illustrates a chromosome region with three genes, with one gene enlarged to show its two parts. Regulatory proteins in the cell bind the regulatory sequence of the gene, determining its on/off status. If "on," the cell makes a working copy of the gene in the form of an RNA molecule called "messenger RNA," or mRNA. The mRNA, once transported out of the nucleus, associates with ribosomes, which proceed to translate the RNA's base sequence into the amino acid sequence of a protein. The DNA of a gene is its stable, inherited form; the mRNA copy made from it is its working form, used or degraded as needed.

During the course of evolution, the DNA sequence of a gene may change. One DNA base, or sometimes more, may be altered (mutated), the result of its slight chemical instability or an effect of some chemical interaction or external radiation. The altered DNA is passed on from one generation to the next. The mutation may have little or no effect, depending on where in the gene it occurred. A mutation may mean that a slightly different protein is made; or the protein may be made by one kind of cell rather than another one; or it may be made later rather than sooner in development; or it may no longer be made by that particular kind of cell at all. The

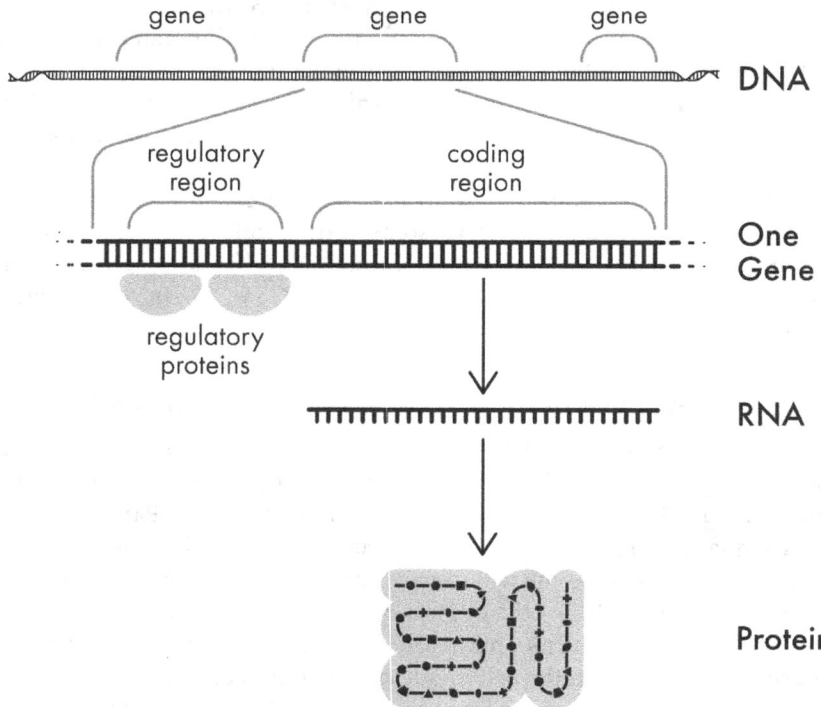

Figure 11. The structure of a gene and how it is used by a cell. The information in the coding region of a gene is copied into an RNA molecule, which the cell then uses to synthesize a protein.

cell's behavior, always dependent on its proteins, could then be altered, like that of a mail boy who has been laid off or assigned a different duty.

The emphasis on genes is justified by their central role in constructing the body through the action of their proteins on the behavior of embryonic cells and also in maintaining the functions of the cells in the adult body. But genes don't determine everything. They have an influence on, but don't completely determine, most of the details of anatomy and physiological function. Note that "identical twins" start out as a single fertilized egg and so have the same set of genes, but they are not really identical in all details—not in fingerprints, not in structural details of the brain and the circulatory system, not in susceptibility to every disease, not in IQ, not in interests or personality. They are similar, but not identical. Other things also play a significant role in how the body's anatomy and physiology and behavior turn out.

We aren't any different from tomato plants in that respect. Yes, a tomato plant is a tomato plant because of its genes. But the kind of tomato plant it turns out to be—its height, its longevity, its flowering time, its yield—also depends on the soil, the temperature, the amount of water and sunlight it receives, the mildew and the insect pests to which it has been exposed.

With humans as with all other animals, chance events during embryonic and fetal development can be important. The chemical constitution of the body, while strongly influenced by how the proteins (encoded by genes) regulate the metabolism of the embryo and adult, is also affected by the amounts and kinds of vitamins, lipids, and amino acids available during embryonic and later development. Particularities of the environment also play a role in contributing to adult health, behavior, temperament, and personality. Otherwise, the quality of parenting, teaching, nutrition, and social interactions—good or bad—wouldn't make any difference. But, as everyone knows, it does.

3

The Embryo

Construction and Continuity

Look in thy glass and tell the face thou viewest
Now is the time that face should form another;
Whose fresh repair if now thou not renewest,
Thou dost beguile the world, unbless some mother.
—William Shakespeare, *Sonnet III* (1609)

Our multicellular bodies do not last. Parts wear out. Cellular and molecular damage accumulates with age, and not all of it gets repaired adequately or accurately. Molecules become entangled or decompose and are not replaced. Genes are occasionally damaged or lost and remain unrepaired and unrestored. But the continuity of the species is assured because at every generation we discard the old worn-out structure and start afresh as a new single cell, which proceeds to build a new body.

How is a new human body made? Not the way an automobile is; the tissues and organs are not made at separate assembly-line stations and joined together at the end with organic screws and collagenous soldering and transmembrane welds. Instead, with a single cell—the fertilized egg, the zygote—the stage is set for the gradual development of all the parts together. The parts are simpler and fewer in number at the beginning, more complex and more numerous as time goes on. They all fit together and function normally at every step along the way to the finished product, the adult body-machine.

The progression from fertilized egg to newborn baby seems so complicated, so mysterious, so planned in advance, and so obviously aimed at producing an adult human being that one might think that pure chemistry could not possibly explain it and that some outside intelligence must be directing the process. But there is no External Agency, no Guiding Hand or Sculptor shaping the embryo as it develops. There is only Molecular Structure ticking away like clockwork, hour by hour, cell division by cell division, driving itself toward birth and adulthood. From egg to adult, the human body *makes itself*, using chemistry and physics. From the beginning the new embryo uses the genes and the molecular toolkits contained in the egg and sperm. With these, along with the environment provided by the mother's reproductive tract and with the assistance of chemical signals between the embryo and the mother, the development of the embryo slowly unfolds. It is the fertilized egg's own internal structure and its content of genes, regulatory proteins, and RNA molecules that allow embryonic development to get started and to proceed along its path. The egg and sperm cells have been prepared in the adult bodies of the previous generation by the normal processes of cell specialization, cell division, DNA replication, and protein synthesis.

Nowhere in the DNA of the fertilized egg will you find a gene for an eye, an eyelash, a hand, a fingernail, a liver, a kidney, a special mathematical ability or musical talent. Nowhere in the zygote is there a miniature version or even any one-to-one representation of adult structures or traits. Instead, what the zygote and early embryo have is genes for proteins that chemically interact with one another and with DNA in such a way as to drive different cells in different directions. The cells mechanically follow the dictates of their protein content, eventually going on to build, purposelessly, bodies that will repeat the whole process and make another round of egg cells and sperm cells.

You could search the internal structure of a fireworks rocket in vain for any miniature version of the flash and bang, the multicolored streamers of light. The ultimate aerial display and noise is absent in the unexploded rocket. There is, to be sure, a *starting* structure which, when the lit match and the fuse are joined, leads to the later flowering of the series of chemical reactions that is the fireworks. The human embryo is that slow explosion of internal chemical structure initiated by the joining of egg and sperm.

I cannot help contemplating this blind process of development without a certain degree of amazement—that a structure built in this manner can eventually feel pleasure, appreciate art and music, and think deep thoughts. Still, why not, if the pure physics of evaporation, condensation, refraction, and reflection can produce a rainbow?

Early Development

Figure 12 illustrates ovulation and fertilization (or "conception"), followed by the embryo's implantation in the wall of the uterus about a week later. Sperm cells swim from the vagina and uterus to the upper end of the Fallopian tube (or oviduct), where one of them may meet an egg (an oocyte) that has recently been released from the ovary. The zygote (the fertilized egg) travels down the oviduct toward the uterus, dividing into smaller and smaller cells as it goes, producing a ball of cells, the morula (a Latin word meaning *mulberry*). These early division stages are called "cleavage stages": to 19th-century embryologists the egg looked as if it was being cleaved into smaller and smaller pieces. With each division a complete double set of genes is distributed to both daughter cells. This continues throughout embryonic and fetal life and on into adulthood; the result is that every cell in the adult body ends up with a complete double set of genes.

The embryo consists of 100 or 200 cells by the time reaches the uterus. It begins secreting enzymes that digest the protecting envelope that has enveloped the embryo up until then. It hatches out—chickens are not the only ones—and the naked embryo, called at this stage a "blastocyst" (another old embryological term from the mid–1800s meaning "growing bladder"), embeds itself in the wall of the waiting uterus.

In the early cleavage stages different cells of the embryo, under the influence of their genes, their local environments, and regulatory molecules

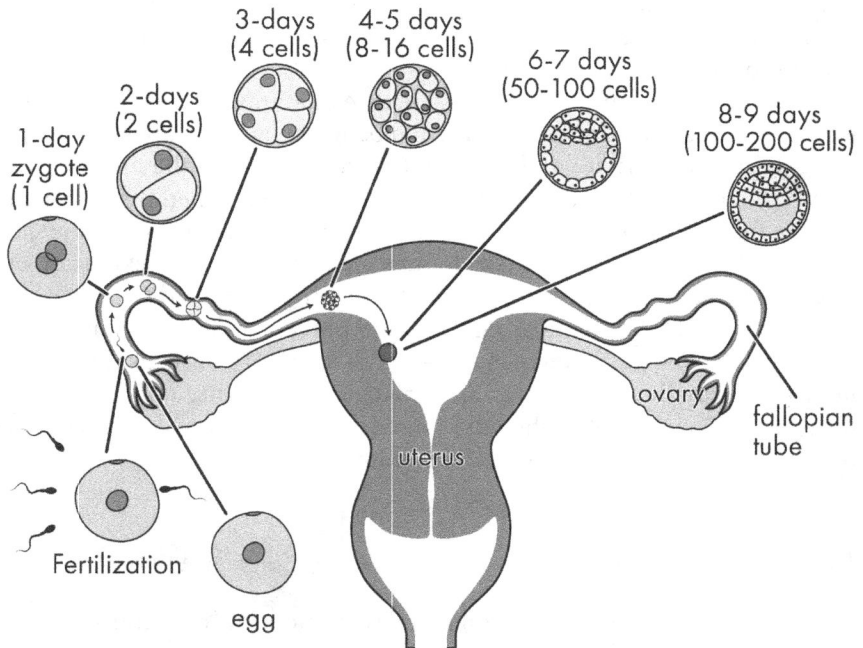

Figure 12. Nine days of human embryonic development, from fertilization to implantation in the wall of the uterus.

they inherited from the egg, are already beginning to behave differently from one another. Before long, cells on the outside of the blastocyst have taken the first steps toward becoming the chorion, the embryonic part of the placenta. Cells on the inside form a layer of cells that partition the blastocyst into two chambers, the amniotic cavity and the blastocoel, the way a tiny coin inside a tiny balloon would divide the balloon into two chambers (Figure 13). The descendant-cells of that partition will become the embryo proper. A few of them will eventually turn into the primordial germ cells, which are destined to generate, many years later, mature sperm cells and egg cells. It is only these cells that will carry humankind into the future. All the rest of the adult body will eventually follow the fate of the placenta. It will die, having played only a supporting role in the continuation of the human line.

How is it that cells all having the same genes come to behave differently? The answer is that different cells use different subsets of all the genes they carry—in the same way that different specialty chefs would read different pages in the same large recipe book. The conditions in the early embryo that bring this about are summarized in Figure 14. First, the egg during its growth in the ovary has been interacting asymmetrically with the surrounding follicle cells. Long before the sperm enters the egg, regulatory molecules have been distributed unequally to different regions of the egg cytoplasm. During cleavage of the zygote, new cells therefore receive different amounts and kinds of these regulatory molecules. This makes a difference as to which genes are used by the different cells (Figure 14, A).

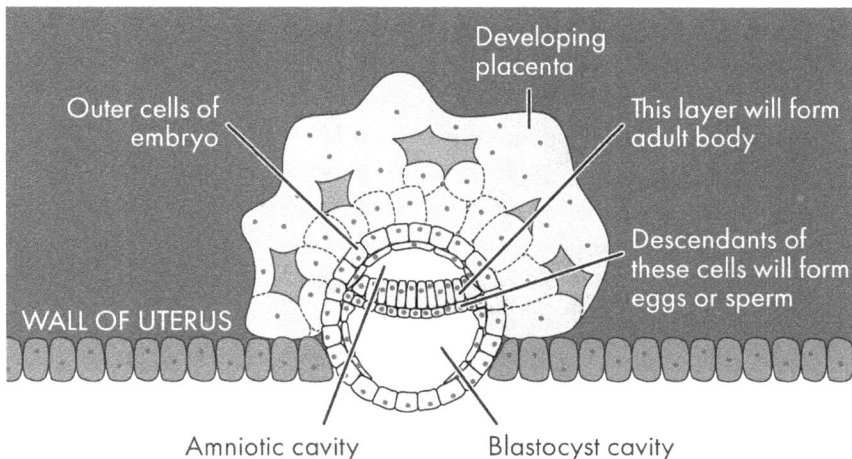

Figure 13. The blastocyst in the second week after fertilization, just after it has implanted itself in the wall of the uterus.

Secondly, after a few cleavage divisions some cells are on the outside of the young embryo, exposed to Fallopian tube fluid, and others find themselves on the inside, imprisoned by the outer cells. The cells in these two locations receive different molecular signals from their environments; the signals are relayed to the cell interiors, and the cells respond by using their genes differently (Figure 14, B). Thirdly, adjacent cells interact. They transmit chemical signals to each other across their membranes, influencing how each uses its genes; genes in different cells can therefore be linked as to how they're used. As the use of some genes in one cell increases, the use of other genes in the other cell automatically declines. The regulatory chemistry of the different cells acts like children at opposite ends of a seesaw (Figure 14, C).

So different cells of the early embryo are driven to using their genes differently. As the embryo grows and the number of its cells increases, the new cells find themselves in an increasing diversity of new environments—adjacent to newly formed spaces and membranes and fluids, or

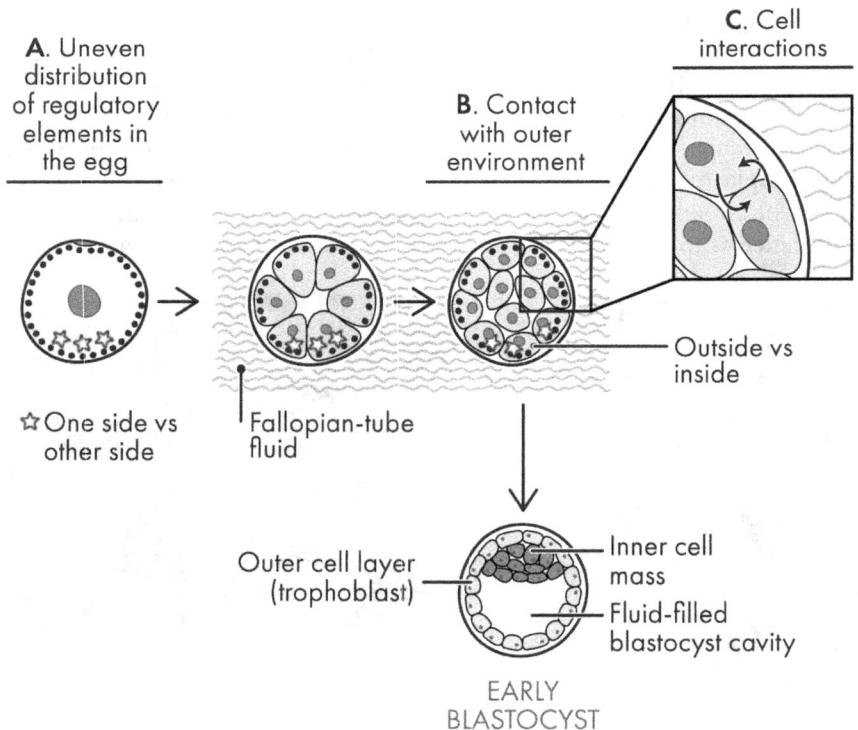

C. Cell interactions

A. Uneven distribution of regulatory elements in the egg

B. Contact with outer environment

☆ One side vs other side

Fallopian-tube fluid

Outside vs inside

Outer cell layer (trophoblast)

Inner cell mass

Fluid-filled blastocyst cavity

EARLY BLASTOCYST

Figure 14. How different cells of the embryo come to use their genes differently: (a) unequal distribution of egg components; (b) different cell locations; (c) cell interactions.

in contact with other cells different from themselves, or coming under the influence of regulatory molecules being secreted by other cells. Just as in the earliest embryo, these external influences cause molecular signals to be transmitted inward to the cell nucleus, bringing new sets of genes into use and propelling the cell in a new direction. As cell division proceeds, cascades of differential gene use flow through different parts of the growing embryo. The gradual unfolding of new sets of proteins and new capabilities eventually results in hundreds of different cell types (see Appendix I) carrying out hundreds of different cellular functions in the adult body.

In trout fishing, there is first the leader, then the worm on the hook, then the cast, then the reeling in, and finally the net. In the development of the embryo, all the events likewise happen in a necessary sequence, earlier ones having to occur before later ones can go forward. The final "terminally differentiated" specialized cell types have all gone through their various stages of development, depending on which genes have been used and what proteins have been made in the cells of the preceding stages.

Continuity

Chickens make eggs, and eggs make chickens. Which came first? In the long chain of evolutionary succession, single cells came before multicellular organisms, so the egg—although it wouldn't have been called that—must have come first.

We take the fusion of an egg cell and a sperm cell as the new beginning of a human life. But it's also an end point, the culmination of the sexual maturation of a body in the previous generation. In fact, life as we experience it now is a continuous process without beginning or end. Once life on Earth had begun, it continued onward in an unending stream—with evolutionary changes and some evolutionary dead ends, to be sure—but what has existed has always been a continuation of what went before.

Making a human embryo and the fetus and adult that follow requires more than the fertilization of an egg. It takes a whole support system provided by the adult bodies of the previous generation. It takes an ovary to grow the egg. It takes a testis to make the sperm cell, and the rest of the male reproductive system to deliver the sperm cell to the right place. It takes the whole female reproductive system to provide a pathway for the sperm to the egg and for the fertilized egg to the uterus. It takes the female system to give the embryo the right fluid environment and the right chemical interactions for its early survival, a uterine wall for implantation, a sophisticated placenta, all the chemical conversations between embryo and mother that take place during pregnancy, the glandular production of milk, and maternal behavior driven by a maternal instinct.

The support systems were put in place, in advance of every fertilization, by the previous generation's development of male and female adults. There is no sharp break, no discontinuity, in the progression of life (Figure 15). Eggs and sperm fuse, new embryos are formed, their cells differentiate into germ cells, gonad, oviduct, uterus, and all the rest, in preparation for the making of yet another round of eggs and sperm, which again fuse, and again new embryos are formed, and again their cells differentiate…

The construction of adult bodies from embryos is the fertilized egg's way of generating more fertilized eggs. Fertilized eggs construct our adult bodies for their own perpetuation. We adults live to serve. In our spare time we amuse ourselves with literature, architecture, agriculture, arts,

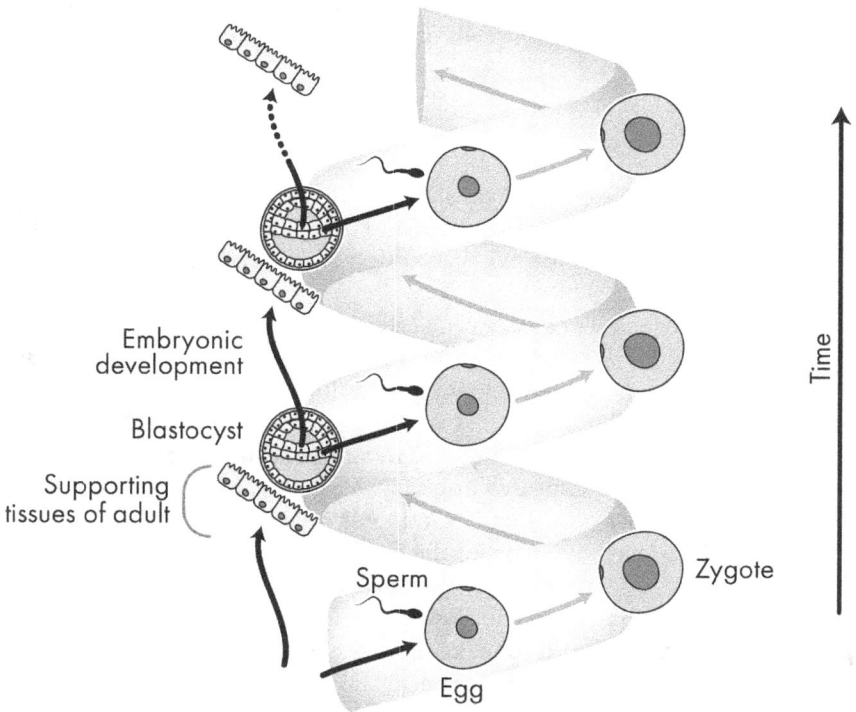

Figure 15. Continuity: the repeating cycle of life, in which each turn of the spiral is another generation. At each generation fertilized eggs constructs adult bodies that prepare environments for the next generation's fertilized eggs in their turn, all the while they are making the next generation's sperm and eggs.

entertainment, sports, dinner parties, science, and philosophy, until we are discarded and the more serious work of fertilized eggs goes on.

The cycles repeat endlessly. Egg and sperm now meet in a favorable environment—the mother's body—because of events that happened in the previous generation. And *those* events depended on the generation before that, and the one before that ... back to the dawn of the human species, you might think, and the eggs and sperm of that time. But the early humans that produced those eggs and sperm did not come from nowhere. They were a link in an unbroken chain connecting them to the eggs and sperm and embryos of more distant, non-human ancestors, and further, all the way back to the first egg cells and sperm cells of the very first multicellular organisms.

4

Evolution:
The Process

Give me a fruitful error any time, full of seeds, bursting with its own corrections.
—Vilfredo Pareto, *Comment on Kepler* (1910)

Edgar Allan Poe wrote,

Science! true daughter of Old Time thou art!
Who alterest all things with thy peering eyes.
Why preyest thou thus upon the poet's heart,
Vulture, whose wings are dull realities?

Is science a dull-winged vulture picking over the bones of the world after destroying its beauty? Does an evolutionary theory explaining the origin of humans from ape-like ancestors diminish our sense of ourselves? Is it cold hearted to look at our species, *Homo sapiens*, as an evolved animal with earthy causes for our basic nature, rather than the spiritual, free-willed creatures we think we are? Are we betraying our non-physical selves by pinning ourselves down like insects for scientific examination?

But understanding my evolutionary origins is to catch a glimpse of my fundamental self. My biological traits, the consequence of evolutionary processes, are as much a part of my inner nature as is anything else about me. Besides, being connected to ancestral ape-like creatures from millions of years ago—from fragmentary fossil remains washed up on evolution's ancient lake shores or buried under the hardened ashes of volcanoes—makes me feel grounded, tied to other forms of life, and securely anchored to the earth and its history. The evolutionary story adds to my appreciation of our human uniqueness as a species. And as far as our present state of knowledge goes, it has the reassuring virtue of being a more-or-less true account rather than a myth.

The Process of Evolution

Charles Darwin, along with Alfred Russel Wallace, described the general *pattern* of evolution, but they didn't know much about the *process*, other than those traits that happened to be beneficial had somehow to be passed on to succeeding generations, where they provided benefits for future survival and reproduction. Ideas about the process try to answer the question, not *what* happened during evolution, but *how* and *why* it happened.

I first heard about evolution when I was 10 or 11 years old. The idea that our teacher presented to our childish minds was this: there is variation in all living things, such as oak trees and sparrows. Only the best ones survive; they are the ones that leave more acorns on the ground or more eggs in the nest. I got it. It was like pebbles passing through a sieve, leaving behind the ones that are the wrong size. As the generations pass, and the tree or bird gives its good traits to its offspring, those types come to represent the whole species. This is not a bad description, even if a little incomplete.

I went out to look at nature's variety. I chose the leaves on the oak tree in my front yard. They were handy, they couldn't possibly be all the same, and they were within reach just waiting to be measured. But then I saw that finding the right kind of variation was harder than I'd thought. The leaves had the wrong kind of variation: they were part of a single individual, and individual leaves don't leave offspring of their own, any more than my individual fingers do. And what about younger *versus* older leaves? I could have sought out leaves from different oak trees, but I wouldn't have known how to do the appropriate statistical analysis. I put the project aside. (My career as a young evolutionary biologist was nipped in the bud. As it eventually turned out, a couple of decades later, I joined a university department where I was surrounded by evolutionary biologists, and the evolutionary viewpoint rubbed off on me.)

A better job of it had been made in 1898 by 35-year-old Hermon Bumpus. In Providence, Rhode Island, early one February morning after a severe winter storm, his Brown University laboratory received 136 frozen English sparrows. Bumpus measured them all; they were variable in size, weight, wing length, and other physical traits. Sixty-four of the chilled birds died, but 72 recovered. On average, the survivors were smaller than the ones that perished. Their slightly smaller size may have enabled them to find better shelter, or maybe it reflected some hidden physiological cold tolerance. In any case, the physical differences were somehow related to the difference in survival.

Bumpus presented his findings later that year in a lecture, "The elimination of the unfit as illustrated by the introduced sparrow, *Passer domesticus*," at the Woods Hole Marine Biological Laboratory in Massachusetts. His concluding remark was "the birds which perished, perished not through accident, but because they did not possess certain structural characters which would have enabled them to withstand the severity of the test imposed by nature; they were eliminated because they were unfit." This appeared to be a concrete example, supported by real data, of what Charles Darwin had been writing about 40 years earlier when he published in *On the Origin of Species by Means of Natural Selection, or the Preservation of Favoured Races in the Struggle for Life*. Bumpus's sparrows showed evolution in action, natural selection at work. The only thing missing, to close the loop of Darwin's logic, was a demonstration that heredity was involved. Without further winter storms, would the next generation's sparrows have been smaller than the ones in the pre-storm population? Would they have survived some later winter storm in greater numbers? Under the circumstances, such a follow-up study would have been impossible to carry out. But Bumpus's measurements nicely demonstrated at least part of Darwin's scheme.

Since then, more thorough studies have been carried out on many plants and animals, both in the laboratory and in the wild. In one investigation, differential survival in populations of a small lizard, *Anolis carolinensis*, was recently found in the southern United States in the summer following an unprecedented cold winter in 2013–2104. The populations of surviving lizards had a greater tolerance for low temperatures than did the lizards in the pre-winter populations, and greater numbers of genes giving cold protection.

A similar kind of "natural experiment," also including evidence for the involvement of genes having survival value, occurred in World War II. On the Eastern Front on September 8, 1941, the siege of Leningrad began. The German Army had cut off all access to the city; the blockade lasted more than two years. When it was over, more than a million people in the city had died, mostly from starvation. Seventy years later Oleg Glotov and his colleagues at the Russian Academy of Medical Sciences in St. Petersburg tracked down over 200 survivors, most of them then in their 80s and 90s. Genetic tests revealed that many of them, more than the population at large, had gene variants that may have provided them an advantage in surviving the winter cold and starvation.

Evolutionary Mechanisms

What were the evolutionary mechanisms that tens of millions of years ago first led to the emergence of mammals, of which we humans are one species, and then to the specializations of the primates (the mammalian order to which we belong), and finally to ancient African apes beginning to walk upright, learning to make tools, becoming omnivores, evolving large brains, and starting to use complex language to communicate?

The characteristic traits of humans did not arise all at once, a whole package delivered to the waiting ancestral ape, suddenly making it human. Most of them probably appeared gradually over millions of years. Furry early mammals progressively began developing in a uterus with the aid of a placenta. The shape of the skull of some mammals was slowly modified to give the forward-facing eyes of primates. Hoofs and claws gradually shifted to become toenails and fingernails. Ape-like anatomy and behavior slid over to become human-like anatomy and behavior. Upright walking and running were fine-tuned bit by bit with slight anatomical modifications of the foot, hand, backbone, pelvis, and skull, followed by modifications of diet, jaws and teeth, and enlargement of the brain. The end result is the layered human being, a creature made of anatomical parts and their correlated behaviors evolving at different times, one feature added on top of another. New elements were formed by modifications of prior ones, but many old features persist. And as new structures and functions arose at each stage, they always had to harmonize with the preceding ones. Evolution is rarely accomplished as steppingstones across a river, with gaps, but as a continuous if irregular path from then to now.

How were these changes brought about? The main bases for evolution seem to be changes in the environment and changes in the body that can be passed on from one generation to the next—changes outside the organism and changes inside the organism. Inside and outside are not really distinct, however. Imagine some organisms gradually sinking deeper and deeper in the ocean. At a few thousand meters down the temperature is near freezing, the pressure is hundreds of times greater than at the surface. There is no light, so grazing on algae is no longer possible. These are environmental changes, you say—but inside the cells the properties of water and membrane lipids are now different, the shapes of protein molecules under great pressure are altered, the rates of chemical reactions are changed, and the kinds of molecules available as food are different—all without any changes in genes. In our evolutionary history there may have been environmental changes that were critical: climate changes, less carbohydrate and more animal fats and proteins in our ancestors' diet. Changes in the DNA of genes would have followed.

We don't have a complete answer, and many details of evolutionary mechanisms, especially in our own species, remain to be clarified. Cosmic accident played its role. In a different solar system, with a different planet tilted differently, at a different distance from its star, and wobbling in a slightly different way, there would have been a different sequence of climate changes and a different sequence of evolutionary events, and no human beings. We can count ourselves to be lucky to be here.

For at least part of the evolutionary process, we do have an outline of an answer: a gene-centered story. It runs as follows. In any population, random gene changes result in genetic variation. Under appropriate environmental conditions, some of the genetic variants give their owners an advantage, however slight, in survival and reproduction. Therefore, those genes are passed on to more and more numerous offspring as one generation succeeds another. Over time, more and more individuals in the population come to carry those advantageous gene variants. The whole population changes, automatically becoming better adapted to the environmental conditions.

More than half a century of studies of genes, DNA, mutation, and genetic variation in natural and laboratory populations of plants and animals has demonstrated the validity of this general view of the process.

A little computer simulation (Figure 16) illustrates the process. On the left in the figure, in generation 1, a change in a gene (a mutation) occurs in one individual. The new form of the gene may contribute to the individual's being slightly better adapted to the environment—perhaps better able to make use of some foodstuff, or better able to escape a predator, or more resistant to some disease—and this might result in a slight advantage in long-term survival or reproduction. At each generation, individuals carrying the new form of the gene will tend to have more than the average number of offspring. Even with the original ones gamely reproducing as usual, over the course of many generations the new ones make better use of resources and living space, and they eventually crowd out the original ones. The population will have gradually shifted over to a new genetic type.

A real-world example of this kind of small-scale evolution is found in the case of resistance to *Plasmodium vivax* in West Africa. *Plasmodium vivax* is one of several species of malarial parasite that invade human red blood cells after the bite of an infected mosquito. This parasite is prevalent in Asia, Latin America, and Africa. (The kind of malaria it causes is somewhat less severe than that caused by the deadly, predominantly African *Plasmodium falciparum*.) The native human population of western Africa has adapted, genetically, to the presence of *Plasmodium vivax*. At some time in the past, in someone, a chance mutation altered a gene in chromosome 1 that encodes a particular red blood cell surface protein, a protein that the parasite uses to enter red blood cells. The mutation prevents developing red blood cells from making the protein, and so the red blood cells are invisible to *Plasmodium vivax*. The result is that descendants of the person in which the mutation first occurred don't get infected by the

A. NEW MUTATION	
Generation Number	Individuals in the Population
1	○○○○○○○○○○○○○○○○○○○○○○○○
2	○○○○○○○○○○○○○○○○○○○○○○○○
3	○○○○○○●○○○○○○○○○○○○○○○○○
4	○○○○○○●●○○○○○○○○○○○○○○○○
5	○○○○●●●●○○○○○○○○○○○○○○○○
6	○○○●●●●●●●○○○○○○○○○○○○○○
7	○●●●●●●●○○○○○○○○○○○○○○○○
8	○●●●●●●●●○○○○○○○○○○○○○○○
9	○○●●●●●●●●○○○○○○○○○○○○○○
10	○●●●●●●●●●●○○○○○○○○○○○○○
11	●●●●●●●●●●●●●○○○○○○○○○○○
12	●●●●●●●●●●●●●●●○○○○○○○○
13	●●●●●●●●●●●●●●●●●○○○○○
14	●●●●●●●●●●●●●●●●●●●○○
15	●●●●●●●●●●●●●●●●●●●●○
16	●●●●●●●●●●●●●●●●●●●●○
17	●●●●●●●●●●●●●●●●●●●●●

B. EXISTING RARE GENE	
Generation Number	Individuals in the Population
1	○○□○○◆○○○○△○○○○○○○○
2	○○□○◆◆◆○○○△△○○○○○○○
3	○○□○◆◆◆◆○○○△△○○○○○○
4	○○□◆◆◆◆◆○○△△○○○○○○
5	○○□◆◆◆◆◆◆○△△○○○○○○
6	○○□◆◆◆◆◆◆◆△△○○○○○○
7	○○◆◆◆◆◆◆◆◆◆△○○○○○○
8	○◆◆◆◆◆◆◆◆◆◆△○○○○○○
9	○◆◆◆◆◆◆◆◆◆◆◆○○○○○○
10	○◆◆◆◆◆◆◆◆◆◆◆◆○○○○○
11	◆◆◆◆◆◆◆◆◆◆◆◆◆○○○○○○○
12	◆◆◆◆◆◆◆◆◆◆◆◆◆◆○○○○○
13	◆◆◆◆◆◆◆◆◆◆◆◆◆◆◆○○○○
14	◆◆◆◆◆◆◆◆◆◆◆◆◆◆◆◆○○○
15	◆◆◆◆◆◆◆◆◆◆◆◆◆◆◆◆◆○○
16	◆◆◆◆◆◆◆◆◆◆◆◆◆◆◆◆◆◆○
17	◆◆◆◆◆◆◆◆◆◆◆◆◆◆◆◆◆◆◆

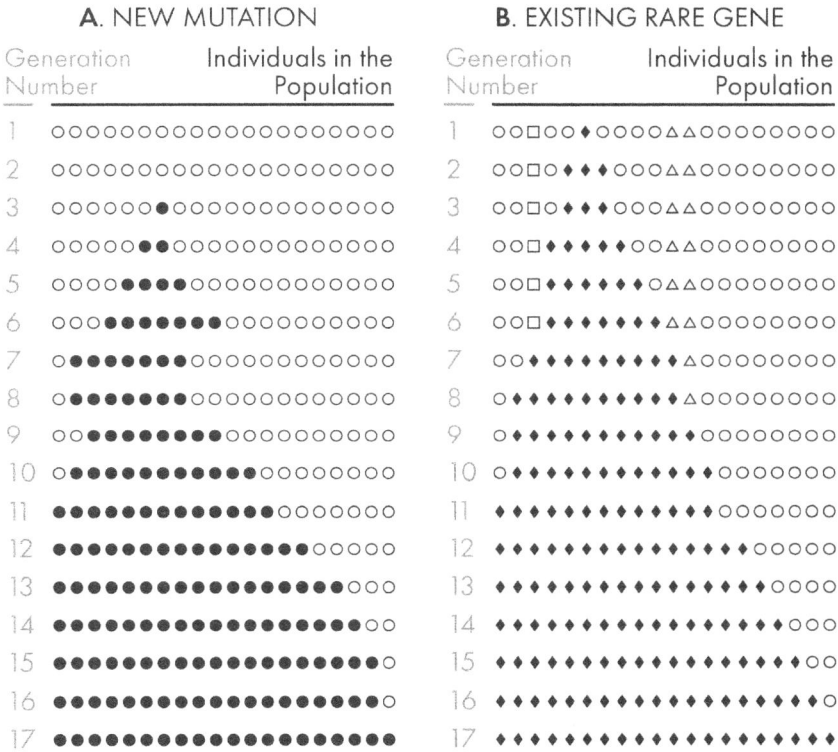

Figure 16. A simple computer simulation showing how evolution by natural selection commonly works. In A, a new gene mutation arises and spreads through the population because of some advantage it provides. In B, the advantageous gene (near the left-hand end of the population) already exists in the population, but under changed environmental conditions it confers a new reproductive advantage that results in its spread through the population, giving the same result.

parasite. Because of that advantage, the mutation, while remaining rare in other parts of the world, has become common in western Africa.

For natural selection to begin to change a population, a new mutation is not always necessary. A mutant form of a gene may already exist among the many gene variants that the population has been harboring for generations. With a change in environmental conditions, that form of the gene may affect fetal development or adult anatomy or physiology in such a way as to give an advantage the gene didn't provide before. This kind of selection—in reality probably more common than the selection of a totally new mutant form of a gene—is illustrated on the right in Figure 16. In the altered environment, offspring carrying a gene that was initially rare in the population (near the middle, in generation 0) now do better at surviving and reproducing. The rare gene has now become the common one in the population. The population has evolved.

This kind of selection, leading to adaptation to the local environment, helps explain some of the geographic differences one sees in human populations around the world. The Inuit of Greenland have a short, stocky body type that conserves body heat better than taller, thinner bodies (conserving heat aids survival), as well as a metabolism that deals efficiently with a seafood-rich diet (the main food available). The people of Africa and of the Indian subcontinent have darker skin (protection against sun damage aids survival), while those living in northern Europe have lighter skin (allowing more sunlight-stimulated production of vitamin D in the body). People of Tibet have genes affecting characteristics of their blood, adapting them to life at high altitudes and low oxygen levels. Northern Europeans have longer and narrower noses compared with those of Africans; it is speculated that they are slightly healthier than they would be otherwise because the longer nasal channel better warms and humidifies the cold dry air before it reaches the lungs. (In evolutionary phrasing, a long, narrow nose has survival value in the north.) The broader noses of native people living closer to the equator in Africa, where *Homo sapiens* first evolved, may have facilitated better air flow and greater dissipation of body heat, especially during times of tracking and hunting.

The selection of genes already present in the founding populations was probably involved in all these examples. Evolution was only a matter of increasing their numbers as the generations passed. These examples illustrate how evolution might have worked, and continue to work, at all times, everywhere: chance mutation by chance mutation over hundreds of generations, and "natural selection" of some forms of genes over others as the environment changed: the genetic differences gradually made a new population better adapted to the new circumstances.

Another evolutionary process is also at work, especially in small populations: genetic drift. With genetic drift there may be gene changes in a population that have nothing to do with natural selection but are the result of pure chance as to who survives and reproduces and who doesn't. A particular form of a gene may provide no selective advantage at all in survival or reproduction, yet still become the predominant form of the gene in the population. A computer-simulation model of genetic drift is shown in Figure 17. Imagine, for the sake of illustration, a small vineyard with 20 green-grape vines. A single red grape, say, arises as a result of a gene mutation. By pure chance (maybe a hailstorm, which kills vines at random), the red grape vine happens to survive and do well for the next three or four generations. But by simple random fluctuations the proportion of red vs. green grapes fluctuates from one generation to the next, in the same way that a series of successive coin tosses will give varying numbers of heads and tails. In one computer run (on the left in the figure), after some fluctuating levels of survival, red grapes have disappeared by the 16th generation. In another computer run (on the right in the figure), after some ups and downs, red grapes have by chance become the predominant type in the vineyard.

In all these cases evolution has been played out in a minor key, without the formation of a new species. But sometimes the accumulated genetic differences might add up to a new species with new traits. This idea would account for the gradual appearance of characteristically human traits in our own evolutionary line: the gradual increase in size over our smaller ape ancestors, the gradual modification of the hands and feet, the gradual adjustment of the whole skeleton for an upright stance, the gradual changes in jaw and teeth and diet, the gradual increase in our brain capacity. The

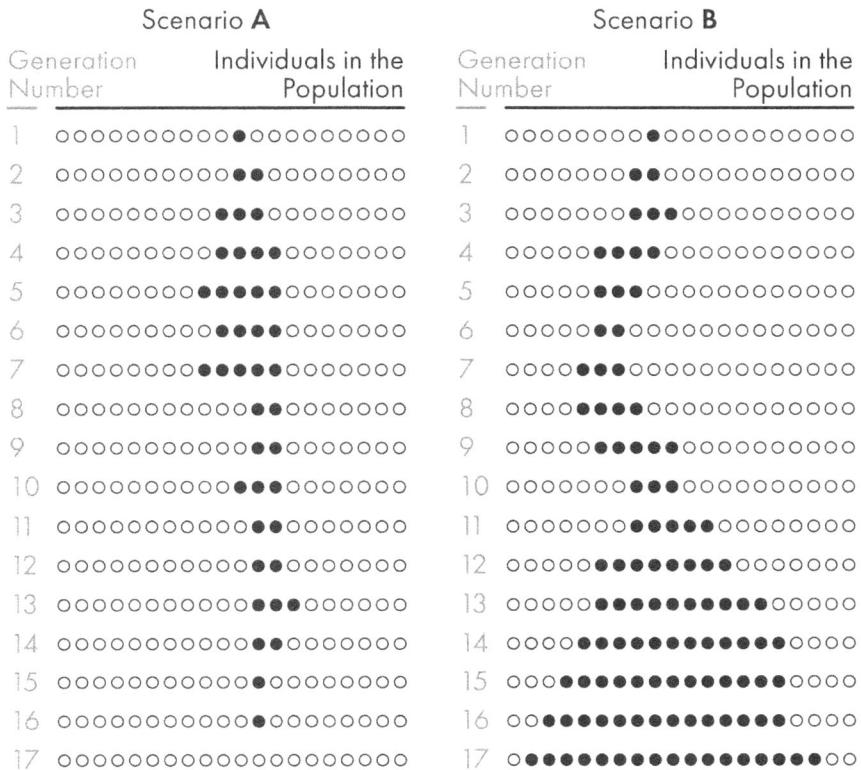

Scenario **A**		Scenario **B**	
Generation Number	Individuals in the Population	Generation Number	Individuals in the Population
1	○○○○○○○○○○●○○○○○○○○○	1	○○○○○○○○●○○○○○○○○○○○
2	○○○○○○○○○○●●○○○○○○○○	2	○○○○○○○●●○○○○○○○○○○○
3	○○○○○○○○○●●●○○○○○○○○	3	○○○○○○○●●●○○○○○○○○○○
4	○○○○○○○○○●●●●○○○○○○○	4	○○○○○●●●●○○○○○○○○○○○
5	○○○○○○○○●●●●●○○○○○○○	5	○○○○○●●●○○○○○○○○○○○○
6	○○○○○○○○○●●●●○○○○○○○	6	○○○○○●●○○○○○○○○○○○○○
7	○○○○○○○○●●●●●○○○○○○○	7	○○○○●●●○○○○○○○○○○○○○
8	○○○○○○○○○○○●●○○○○○○○	8	○○○○●●●●○○○○○○○○○○○○
9	○○○○○○○○○○●●○○○○○○○○	9	○○○○○●●●●○○○○○○○○○○○
10	○○○○○○○○○●●●○○○○○○○○	10	○○○○○○○●●●○○○○○○○○○○
11	○○○○○○○○○○●●○○○○○○○○	11	○○○○○○○●●●●●○○○○○○○○
12	○○○○○○○○○○●●○○○○○○○○	12	○○○○○●●●●●●●○○○○○○○○
13	○○○○○○○○○●●●○○○○○○○○	13	○○○○○●●●●●●●●●○○○○○○
14	○○○○○○○○○○●●○○○○○○○○	14	○○○○●●●●●●●●●●●○○○○○
15	○○○○○○○○○○●○○○○○○○○○	15	○○○●●●●●●●●●●●●○○○○○
16	○○○○○○○○○○●○○○○○○○○○	16	○○●●●●●●●●●●●●●●○○○○
17	○○○○○○○○○○○○○○○○○○○○	17	○●●●●●●●●●●●●●●●●○○○

Figure 17. A simple computer simulation showing genetic drift, with pure chance determining different outcomes.

same processes that are occurring in microevolution, over longer time spans might result in the accumulation of enough gene differences to give a new population that no longer interbreeds with the original population. The new population would then be deserving of a new species name.

It is harder to get concrete evidence for this kind of evolution, called *macro*evolution, involving not just new species but whole new families of species. The timescales involved are not thousands but millions of years. Original ancestral populations have often been extinct for millions of years or more and are more difficult to identify. Nevertheless, anatomical and genetic relations of species and of larger species-groups fit the patterns expected if the gradual accumulation of genes giving new adaptations is responsible for new species formation.

The two species *Homo sapiens* (humans) and *Pan troglodytes* (chimpanzees) are very similar genetically. We probably had a common ancestor, some kind of human-like ape that lived in Africa about seven million years ago. Since then, chimps and humans have evolved in different environments—the chimps (like our probable evolutionary ancestors) mostly in forests, and humans in more open woodlands and savannas. The foot of the two species reflects their different uses (Figure 18). Chimps that a couple of centuries ago were called "quadrumanous," or "four-handed," use their feet not only for traveling along the ground but also for climbing and grabbing things. Humans use theirs for walking and running (hence the alignment of the big toe), and sometimes for kicking, stamping, or dancing. Fossil evidence of African great apes from three to five million years

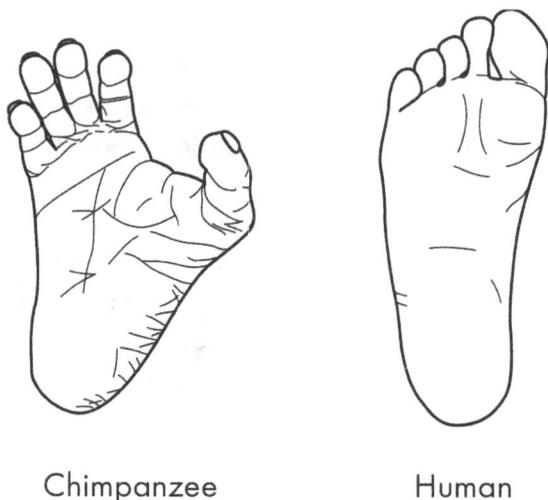

Chimpanzee   Human

Figure 18. The chimpanzee foot and the human foot.

ago indicates that they probably spent a fair amount of time on the ground and had feet in many ways intermediate between those of modern chimps and humans. It looks, therefore, as if the modern human foot evolved by means of small, successive gene changes that gradually molded the embryological development of an ancient ape ancestor's foot into the modern human form.

In this case the particular genes and proteins involved in changing the human foot to its present form have not been identified. There are nevertheless a few hundred specific gene differences between chimps and humans, and some (many?) of them, although we don't know which ones, might have been the ones that created the initial irreversible non-interbreeding divide between chimps and humans. But at some point six or seven million years ago, our ancestors stopped mating with the ancestors of chimpanzees. A new species had arisen on the savanna; one of its evolutionary descendants would be *Homo sapiens*.

Adaptation and the Randomness and Purposelessness of Evolution

A short 1960 film called *Day of the Painter* depicted a painter who early one morning walked out onto a mud flat at low tide and pegged down the four corners of a large canvas. Back up on the pier, he assembled an assortment of cans of paint, which he proceeded to slosh down onto the canvas below. He cut the canvas into a few dozen 2' × 3' rectangles, which he carried back up to the pier and tacked onto wooden frames. Around noon the tide came in and the Tourists arrived. By the time the sun had sunk down to the horizon he had sold a few paintings, generating a small profit for the day's work. The other paintings he tossed into the water, and they were carried out to sea on the outgoing tide.

The process of evolution is like this painter's method. Its various canvases are produced by a random process, and it has no particular aim. What is saved is merely what happens to be chosen by the Tourists of the moment. Some species, their genes, and the structures their proteins create, are preserved as circumstances happen to dictate. The others float away into oblivion.

Consider the giant anteater. It has strong foreclaws for digging out ant nests and tearing open termite mounds, an elongated snout, a two-foot-long slender, sticky, raspy tongue for collecting the insects from their tunnels, and a stomach with rough, hardened folds for grinding up the insects. It seems well designed for how it makes its living.

Consider that relative of the shark, the stingray. Its body is flattened, and its wing-like pectoral fins extend out sideways for cruising along shallow sandy ocean bottoms in search of food. Its eyes are on the upper side so that it can still look out for passing prey when it partially buries itself in the sand. Just behind the eyes on the upper side are special openings for taking in sand-free water. The mouth is on the ventral side. Also on the ventral side are numerous small organs for sensing faint electrical signals that accompany the muscular movements of its prey, buried clams and other mollusks. These the stingray uncovers and then crushes with special corrugated, flattened teeth. The stingray seems admirably well designed for its way of living.

Consider a virus—not quite alive but nevertheless capable of inserting itself into our cells and making thousands of copies of itself. The virus particle has a protein coat containing special proteins for attaching to the surface of cells in our body. Once inside the cell, the virus's genes are used for making the proteins that are responsible for replicating more viral genes, for making more coat proteins, and for helping assemble all the parts of new virus particles.

Consider the human body. It has sweat glands for cooling the skin, tears for lubricating the eye, a forward-pointing big toe for pushing off on a run, teeth for chewing food, digestive juices for digesting it, a heart for circulating the blood, clotting factors for preventing too much blood loss after an injury, an immune system for protecting the body against bacteria and viruses, eyes in order to see, ears in order to hear, and so on. The human body seems well designed for human life.

The key words in all the foregoing descriptions are *for* and *in order to*. That phrase, and the insidious, tendentious little word *for*, implies *purpose*, which implies a *goal*, a *design*, and advance planning to achieve the goal. The anteater has strong claws for breaking open termite mounds, the stingray has electric organs on its underside for detecting prey buried in the sand, the virus has a protein coat for invading its target cells, humans have an opposable thumb for grasping tools and other objects.

It's entirely natural to think in these terms. That's the way we human beings deal with the world and operate in it. We make can openers for opening cans, doors for opening and closing, umbrellas for keeping off the rain, lamps for giving light, hammers for pounding nails, vacuum cleaners for sucking up dust, stoves for cooking, and so on—design and purpose everywhere. Why should it not exist also in the natural world? But it doesn't rain in order for grass to grow, nor for people to put up umbrellas. It rains, and grass grows and umbrellas open. Rain has consequences, but not purpose.

Nevertheless, it's difficult not to think in terms of purpose when thinking about the living world: the eggshell is to protect the growing chick, milk is for nourishing the lion cub, red blood cells are for carrying oxygen, estrogen is produced by the ovaries in order to regulate the female reproductive system. Even evolutionary biologists continue to think in such terms—although strictly speaking they shouldn't. To see the living world as just part of the vast automatic machine that is the Universe requires a different way of thinking, an altered vision opened up by Charles Darwin.

Two man-made devices, the pocket watch and the mousetrap, are frequently used for comparison with the organisms of the biological world and how they might have come into being. William Paley, English clergyman and philosopher, famously wrote in 1802 as follows (in which the "artificer" is, of course, God):

> Suppose I found a watch upon the ground ... when we come to inspect the watch, we perceive—what we could not discover in [a] stone—that its several parts are framed and put together for a purpose ... the inference we think is inevitable, that the watch must have had a maker—that there must have existed, at some time and at some place or other, an artificer or artificers who formed it for the purpose which we find it actually to answer, who comprehended its construction and designed its use....
>
> The contrivances of nature surpass the contrivances of art, in the complexity, subtlety, and curiosity of the mechanism; and ... [are not less] suited to their office than are the most perfect production of human ingenuity.

The household mousetrap (Figure 19) consists of a base platform, a spring, a catch where the bait is placed, a catch lever or hold-down bar, and a hammer bar. It works only when all the parts are put together; anything less is useless for killing mice. An owl also consists of an assembly of parts that together make a good mouse-killing machine: silent wings, acute hearing, keen eyesight, a beak, and sharp claws. Could a creator not have also the assembled the parts of an owl, like a mousetrap, perhaps all in one go? There's a superficial similarity as to the coordination of parts that we see in these two mouse-killing machines, but they achieved their form and function in totally different ways. People had used flat pieces of wood, springs, and thin metal bars and other differently shaped pieces of metal for different purposes for a long time. Then in 1894 William Hooker had a vision, hit upon a design, and put the different parts together to make what is now the standard household mousetrap.

By contrast, the owl developed slowly and gradually over the course of several million years, small step by small step. The different parts of the modern owl were not just lying around waiting to be put to a new use. Their evolutionary ancestors were birds that had hearing and eyesight that were probably not quite as good as those of the modern owl. They flew not quite as silently and had claws and beaks not quite as well adapted to detecting and catching small rodents. While ancestral rodents were becoming

Hold-down bar

Platform

Hammer

Catch

Spring

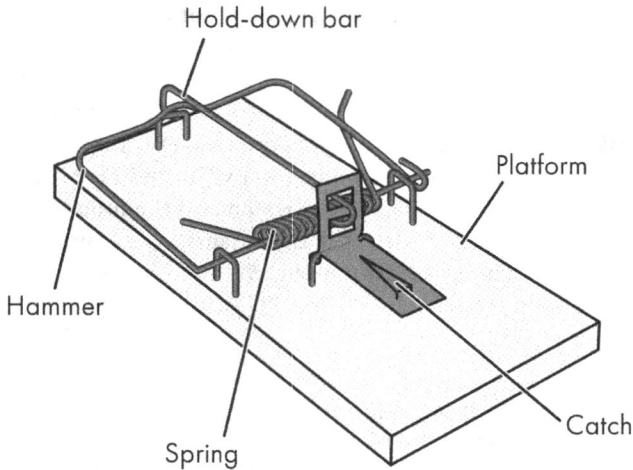

Figure 19. The common household mouse trap.

smaller, quicker, more cautious and secretive, and surviving in higher numbers by avoiding predatory birds, ancestral owls on their side were evolving better eyesight, better hearing, more silent flight, better claws and beaks—and catching more mice, eating better, and having more owlet offspring. In this way owl and mouse populations involuntarily inched their way toward a greater degree of precision in their respective lifestyles, balancing antagonistic tendencies driven by the long-term outcomes of better survival and reproduction, like two poker players each upping the ante for a potentially better payoff. But in the never-ending game of life there is never a final call: owls and mice may continue to coexist, uneasily. Each has evolved and adapted gradually, the mouse to its environment, which includes owls, and the owl to its environment, which includes mice. This is a classic case of—choose your term—an evolutionary arms race, evolutionary feedback, or antagonistic coevolution.

This view of how a species achieves apparently near-perfect form and function adaptations to its environment provides a better explanation than any other. The mechanisms of evolution are blind, without purpose and without any vision of future benefits. A few of all the random gene mutations that occur may happen to sharpen a few mouse ears, soften a few owl feathers, make an immune system a little faster to respond to a virus—and there will automatically be an increase in the number of offspring carrying those genes. The number of those genes increases, and the population changes.

This idea can in theory explain everything about all the marvelous adaptions we see in nature—about how organisms come to be adapted to their environment, to survive, to reproduce, often in remarkable ways. We humans are part of nature too, so why shouldn't that idea also apply to us and our way of being? This was Darwin's idea, that evolution is purposeless, without goals. It was never aimed at producing the human species. Ancestral apes were pushed from behind, as it were, by random mutations in particular environmental circumstances. Some of the apes that by chance happened to have certain genetic variants had more offspring than others, and so their descendants evolved in a certain direction. Under other circumstances those ancient apes might have experienced different climate shifts and different ecological changes, and as a consequence they might have evolved in some other direction. Some other kind of primate might have evolved, not us. We might never have existed except as one among many theoretical but unrealized possibilities.

"For the good of the species"

Related to the notion that a species evolves toward a goal of adapting to its environment is the fallacious belief that the species' traits exist "for the good of the species"—as if *The Species* were some commanding, all-knowing entity that looked out for its own welfare, directing how its members' traits were to be formed and were to function. This subtle and seductive phrase *for the good of the species*, applied to humans, has bewitched many a pre–Darwinian philosopher. In 1819 Arthur Schopenhauer wrote the following in "The Metaphysics of the Love of the Sexes":

> Everyone will decidedly prefer and ardently desire the most beautiful individuals; in other words, those in whom the character of the species is most purely and strongly marked.... The delusive ecstasy that seizes a man at the sight of a woman whose beauty is suited to him, and pictures to him a union with her as the highest good, is just the sense of the species.... The maintenance of the type of the species rests on this decided inclination to beauty; hence it acts with such great power.... Therefore, what here guides man is really an instinct directed to what is best for the species, whereas man himself imagines he is seeking merely a heightening of his own pleasure. In fact, we have in this an instructive explanation of the inner nature of all instinct, which, as here, almost always sets the individual in motion for the good of the species.

This mistaken idea has the virtue of being clearly enunciated. Instinct—an innate, gene-based pattern of behavior—does indeed set the two individuals in motion, the man toward the beautiful woman and the woman toward the right man. But the attraction of the sexes it is not for the good of the *species*. It exists because it results in the individuals' own successful procreation. In making their offspring the male and the female are creating vessels that carry copies of those same genes that underlie their mutual attraction. Those genes are part of the mechanism that serves the genes' own continued propagation. The instinct didn't arise because it's what was best for the species. It arose because it's a mechanism by which the underlying genes for successful reproduction can be transmitted. One could say that because the species endures, the species as a whole benefits. But that's just a downstream consequence of the mechanism of individual survival and reproduction. It's not *for the species* that the mating instincts arose; they arose because they facilitated their own propagation—in those gene vessels, namely the offspring of mating individuals. The mating instincts are the lifelines that genes throw across the ravine to pull themselves into the next generation.

We don't yet have the full story about human evolution. We don't have all the fossils we'd like, and for the fossils we do have, we're not sure which ones might represent our true evolutionary ancestors rather than just side branches on our evolutionary tree. Uncertain also are the details of the ecological shifts that underlay the evolution of *Homo sapiens*. It's probable that a change in climate in Africa a few million years ago, with the shift from forest to savanna, had a lot to do with the evolution of upright walking and running, and the attendant evolution of our hands and our ability to make tools. These developments permitted the addition of meat to our ancestors' diet and allowed the evolution of larger brains. But we know little of the environmental factors that might have led to the evolution of our ability to plan ahead, to interact socially with one another in highly subtle ways, and to communicate by means of sophisticated language.

However we define being human, the change to *Homo sapiens* didn't happen all at once. There was no blinding flash, no sudden leap across an evolutionary ravine, no abrupt transformation from non-human to human. Our existence depends on there having first evolved some ancient, sophisticated ape with the anatomy, brain, and behavior already poised to slide imperceptibly into humanness. All that was required were some minor modifications of that pre-existing ape, the way you might add a cushion or a headrest to an armchair whose basic design is already set.

Are We Still Evolving?

It might be thought that human evolution is all in the past now that it has reached its supreme goal and has led us to our present state of near perfection. It might be thought, also, that the rise of technology and modern medicine means that previously "unfit" individuals, those that couldn't see well enough to notice the approaching lion or run fast enough to escape it, can now survive and be able to pass on their genes to their children just as well as anybody else—as if we have stopped evolution in its tracks, or at least have taken charge of it ourselves.

We have good reason to believe that this thinking is not correct and that we are still evolving.

Evolution is principally the result of having a genetic constitution that under particular environmental circumstances enables some to live longer and have more offspring than others. There will always be a statistical tendency for some people, because of their genes, to be slightly healthier or more prolific than other people. These people would pass on their genes to more children than other people could. It's this kind of evolutionary change on a local scale that accounts for the genetic adaptations—different physical and physiological traits—that have evolved in different populations ("races") in different parts of the world. The rate of such genetic change is slow, on the scale of many thousands of years. It has occurred, for example, in people of the high-altitude terrains of Tibet and the Andes, in people in the cold latitudes above the Arctic circle, and in the light-skinned, lactose-tolerant people of Europe thousands of years ago.

Only if some genes provide advantages in baby-making worldwide regardless of local environments, and only if the changed environmental circumstances favoring some genes over others last hundreds of generations (for example, with permanent climate change), would the genes in question likely increase in number in the long run and change the human species as a whole.

Nevertheless, the general principle of differential survival and reproduction must always be true, even though these days nearsightedness and lions are rarely important factors in who survives to have offspring and who doesn't. Different factors must be operating now that we have civilization, dense populations, agriculture, better medicine, widespread travel and more extensive mixing of genes.

The genes that might be expected to be participating in any future evolutionary change in our species—that is, making a difference in how many children are being produced—are genes that result in (a) a slightly better immune system or give resistance to specific highly person-to-person transmissible viral diseases such as AIDS and Covid-19, (b) resistance to diseases transmitted from domestic animals, (c) reduction in blood pressure and the risk of heart disease in spite of high fat and high salt diets, (d) modification of physiological traits so as to allow a better chance of surviving periods of drought and starvation, (e) extension of life by a few years, enabling greater fertility and additional care for grandchildren, or (f) alteration of our brains to make us more sociable. A few observations align with some of these possibilities, but it's hard to get solid data because evolutionary change is such a long, slow, subtle process.

All the elements of the evolutionary process are currently in place, as they have always been. Evolution must be occurring, but the process is too slow for us to see at what rate and in what direction. Theoretically the accumulation of gene changes in the face of major long-term environmental changes could lead to enough genetic change that we would no longer be *Homo sapiens*. That would be evolution on a grander scale and would probably take a million years.

5

Evolution:
The Pattern

The advance of science is not comparable to the changes of a city, where old edifices are pitilessly torn down to give place to new, but to the continuous evolution of zoologic types which develop ceaselessly and end by becoming unrecognizable to the common sight, but where an expert eye finds always traces of the prior work of past centuries.
—Jules Henri Poincaré, *Valeur de la Science* (1905)

The *pattern* of evolution answers the questions, *what happened, and when*? Two main sources of information allow us to reconstruct the course of our evolutionary past: (i) comparisons of fossils whose ages have been determined by physical and chemical methods and (ii) the pattern of genetic (DNA) relatedness among living species. These comparative genetic-relatedness studies start with our closest relatives the (other) great apes—chimpanzees, gorillas, and orangutans—and then are expanded to include the lesser apes (gibbons), then monkeys, then all primates, then all mammals, and so on, including ever-widening circles of animal life. The inescapable conclusion from both the fossils and the DNA data is that humans evolved from extinct apes. Those apes themselves evolved from more ancient primates, which looked a little like the modern lemurs of Madagascar. Those ancient primates evolved from rodent-like animals now long extinct. And those rodent-like animals evolved from even more ancient reptile-like animals.

Darwin's Tree

A framework for the pattern of evolution and the evidence for it was provided by Charles Darwin in his *The Origin of Species* (first edition, 1859).

The idea was independently formulated around the same time by Alfred Russel Wallace, who based his evolutionary proposal on his explorations of the Amazon and the Malaysian archipelago. The theory of evolution is sometimes called the Darwin–Wallace theory.

A page from one of Darwin's notebooks, dated 1837 when he was 28, is shown on the left side of Figure 20 below. It shows new species branching out from old ones. This accounts for the patterns of relationships among different species. There is a species at "A" with two closely related species, another species at "B" with another two related species; and the same for "C" and "D." The species groups at "C" and "D" are more closely related to each other than either group is to "A" because they arose from a single recent ancestor. One would have to go back further in time (the lower-most branch point) to find the ancestor that "A," "B," and "C" all had in common. Many species have gone extinct; some are still with us. This is the common pattern, confirmed by modern genetic studies: a separate population is formed from a pre-existing species, after which it begins following its own separate evolutionary pathway.

An ancestral species almost always gives rise to just one new species rather than into two or more simultaneously as Darwin's tree has it. The reason, as we now recognize, is that new species commonly arise from just one isolated population of the ancestral species, not from two or more simultaneously. Darwin's tree could therefore be updated and simplified by pruning it. The right panel of Figure 20 shows the tree Darwin might have drawn if he had diagrammed the evolution of the great apes and had had the information we now have. He would have known, and his diagram would have shown, that chimpanzees are not our evolutionary ancestors. Chimps have

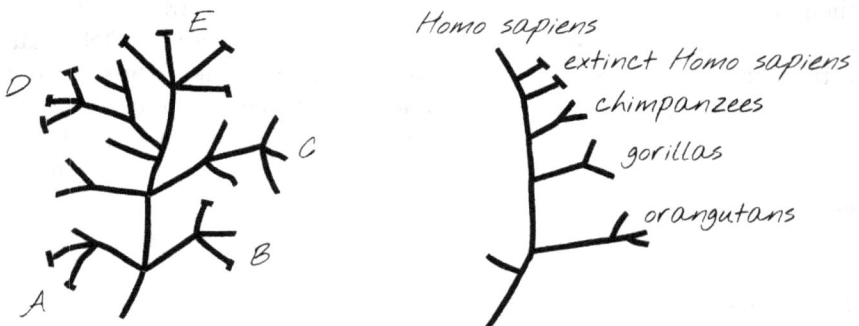

Figure 20. Left, an evolutionary tree similar to one Charles Darwin drew in his 1837 notebook. Different species or species-groups arise as successive branchings from pre-existing ancestors. Right, a more modern evolutionary tree of the great apes preserves Darwin's basic idea.

their own evolutionary history, separate from ours, although humans and chimpanzees had a common ancestor several million years ago.

It may seem odd and arbitrary that life on Earth should be constructed in such a way that a chimpanzee rambling through a rainforest in Tanzania last Saturday and I sitting at my desk in my office today should have in common some ancestral apes that lived seven million years ago. We two are now separated by ways of living and by continents; why should it not always have been so? Why should we not have come into being by separate events of creation, however that might have happened? The answer is that it's just the working out on a larger scale of the same law of reproduction that gives my cousin and me some similarities because we have grandparents in common.

With more and more branches of life slowly coming into being all the time, why hasn't the world filled up to bursting with species? The answer is the same as the answer to why the world doesn't fill up with my first, second, and third cousins: some die, to be replaced by their descendants. It is the same with species: some go extinct and are replaced by others.

There is the added element of *differential* reproduction. My cousin has more children than I do, no doubt by chance, but also maybe because of something innate—that is, genes. Therefore, in the future it is likely that it will be his descendants, not mine, who will be roaming the streets of our town. So it is with the species of the Earth.

In his *Autobiography* Darwin wrote, "[The] problem is the tendency in organic beings descended from the same stock to diverge in character as they become modified. That they have diverged greatly is obvious from the manner in which species of all kinds can be classed under genera, genera under families, families under sub-orders, and so forth; and I can remember the very spot in the road, whilst in my carriage, when to my joy the solution occurred to me ... that the modified offspring of all dominant and increasing forms tend to become adapted to many and highly diversified places in the economy of nature."

Evolutionary Genetics

One source of information about the evolutionary past is the present. Comparison of the sequences of genes and proteins of the living great apes—orangutans, gorillas, chimpanzees, and humans (a specialized great ape)—reveals a non-random pattern of gene changes. The general pattern, seen in DNA and protein sequences throughout the animal and plant worlds, is illustrated in the table below. In the table the letters a-g stand for particular forms of four different genes or proteins, 1, 2, 3, and 4. In gene

1, all four species have the same sequence, here designated "a"; in gene 2, the human gene is different from the other three; in gene 3, chimps and humans are the same but different from orangutans and gorillas; in gene 4, orangutans are different from all the others.

Gene	1 2 3 4
Orangutan	a-b-c-d
Gorilla	a-b-c-g
Chimpanzee	a-b-f-g
Human	a-e-f-g

This pattern of gene or protein changes (mutations) is not haphazard. It reflects an evolutionary history. While all four species share the sequence "a," the gorilla, chimp, and human share "g," so these three species are more closely related to one another than any one of them is to the orangutan. The chimp and the human also share "f," so they are more closely related to each other than either is to the gorilla or the orangutan. Finally, on the human twig of this evolutionary branch, a recent mutation changed "b" to "e," making humans distinctive for gene 2.

Such patterns make sense as being the consequence of a series of successive mutations. This is shown in Figure 21. At an early stage, sequence "d" mutated to "g," and all the evolutionary descendants from then on share "g." Later, sequence "c" changed to "f," and the two new species on that branch had "f," inherited from their common ancestor. Later, sequence "b" changed to "e."

This is the universal evolutionary pattern—each new species changes or adds on to what preceded it, rather than independently coming up with its own sequences. It is one of strongest pieces of evidence for evolution we have. The same kind of pattern, based just on the analysis of the DNA and proteins of living species, is typical for all plants and animals. It reveals their evolutionary relationships and allows the construction of an evolutionary tree of life. Darwin was right: such patterns must have been generated by ancestral populations successively splitting into new species. What we have today are their living descendants. We humans are one of those.

Fossils

Evolutionary trees constructed in the way just described show the pattern of evolution, but they have no timescale and no geography. How long ago did the branch that led to modern orangutans diverge from the gorilla-chimp-human branch? How long ago did the branch leading to

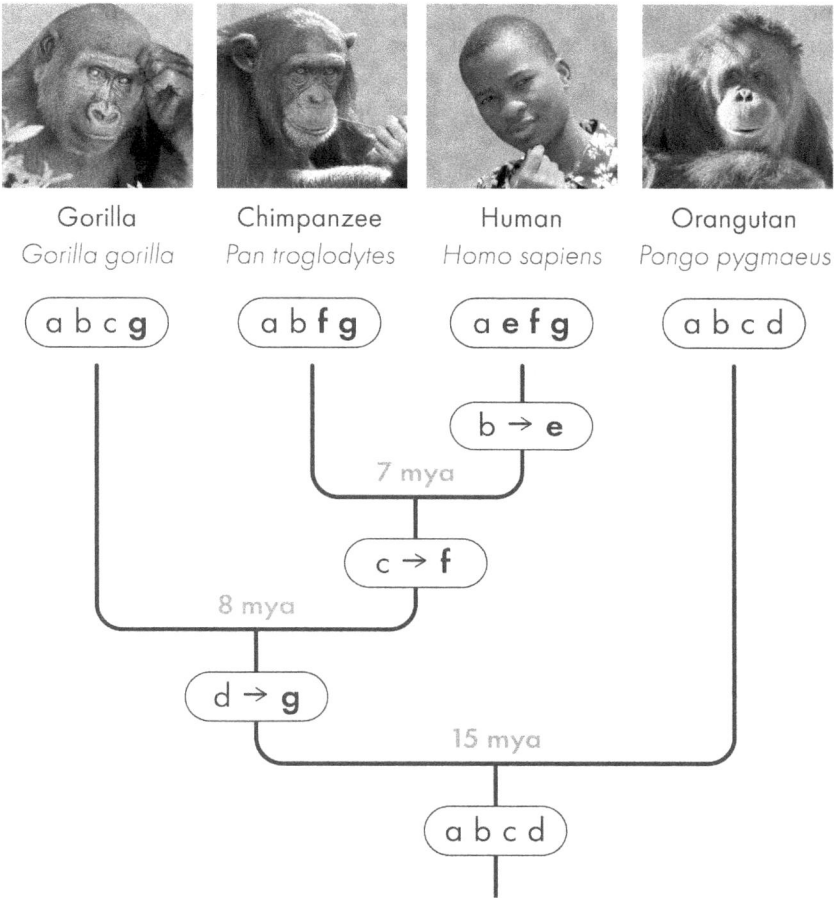

Gorilla
Gorilla gorilla

Chimpanzee
Pan troglodytes

Human
Homo sapiens

Orangutan
Pongo pygmaeus

Figure 21. The great-ape evolutionary tree, reconstructed from the pattern of changes in genes (or proteins): d to g, c to f, and b to e. Estimates of the time of different evolutionary branching events are given as millions of years ago (mya, millions of years ago).

modern humans separate from the branch that led to modern chimpanzees? When did the ape ancestor that was common to both modern chimps and modern humans live? And where in the world did these evolutionary events occur? The timescale for evolutionary trees, guessed at from estimates of rates of genetic change, can be reinforced and put in its geographical context with the help of fossils.

Fossils—buried skeletal or other remains (the word is from the Latin *fossus*, "dug up")—of extinct animal species are notoriously hard to find, almost always incomplete, and their affiliations with each other and with modern species are often difficult to decipher. Nevertheless, their anatomical similarities and differences and their geographic locations can give clues about their relationships. More importantly, fossils can be dated. Therefore, to the extent that they can be related to one another and to living species, they provide a more certain evolutionary timescale than can be given by mere genetic-relatedness patterns of living species.

Fossils can be dated by physical and chemical methods. With radiocarbon dating, use is made of carbon-14, a radioactive isotope of carbon that is present in small, fairly constant amounts in the atmosphere and in living tissues. The ages of carbon-containing fossils can be estimated by how much carbon-14 has been lost by radioactive decay since the death of the animal or plant. On geological timescales the radioactive decay of carbon-14 is relatively fast, and fossil remains that are very old have too little carbon-14 to be measured. For fossils older than about 50,000 years, other methods based on physical or chemical processes with slower rates of change are necessary.

One widely used method for dating rocks and minerals associated with fossils makes use of their content of crystalline quartz, which is common in granite, feldspar, lava, shale, and sandstone. Very slowly over long periods of time, quartz crystals gradually accumulate imperfections caused by natural radiation in their environment, causing changes in the distribution of the crystal's electrons. Stimulated by light or heat in the laboratory, the electrons release their energy in the form of a faint luminescent glow, which can be measured. The amount of luminescence is proportional to the number of crystal imperfections, and therefore to the amount of time that has passed since the rock or sand was last exposed to sunlight or high heat. The method is useful for dating fossils hundreds of thousands of years old.

Another method for dating even older rocks and sand associated with fossils is based on the very slow rate of radioactive decay of potassium-40 (^{40}K), which is present in about one one-hundredth of 1 percent of all the potassium in the rocks. Radioactive ^{40}K decays at timescales in the billions of years, and so can be used to date very old fossils. What is usually measured, rather than the loss of ^{40}K, is the accumulation of one of its decay products, the gas argon-40 (^{40}Ar), which remains trapped until the sand grains or fragments of rock are heated or irradiated in the laboratory. The method is called the potassium-argon method ($^{40}K/^{40}Ar$). A modified version of it is the argon-argon method ($^{39}Ar/^{40}Ar$), which measures the base amount of potassium indirectly by converting it into a different isotope of argon. These methods are useful for determining the ages of fossils that are millions of years old. Assuming that the fossil hasn't been separated from the nearby rock by earthquakes or flowing rivers or carnivores carrying bones from one place to another, or some other kind of disruption of the original site of the death of the ape or other animal, the method is typically accurate to within about 10 percent or so (a fossil dated at 100 million years might really be 90 or 110 million years old), which is accurate enough for most purposes.

Other methods make use of the decay of other radioactive elements on other timescales, or associations with rock strata showing reversals of earth's magnetic field, or biostratigraphy (associations with biological materials such as fossil animal or plant species whose ages have been determined in other locations).

Fossils that might be the remains of ancestors of all the great apes past and present are all from Southeast Asia. That is where those ancient apes probably first appeared during the early Miocene epoch around 20 million years ago (see Figure 22). Fossils of later Miocene apes from 10 to 15 million years ago have been found from India to Europe and northern Africa. We have almost no fossils of gorilla or chimpanzee ancestors later than those Miocene apes (African jungles are not favorable for fossil preservation), but from the present-day distribution of gorillas and chimps and from human fossils (some of which may have been our evolutionary ancestors), central and eastern Africa was the likely location of the apes that were the ancestors of all the African great apes—gorillas, chimpanzees, and humans. A fair number of related fossils have also been found in southeast Asia, showing that the evolution of large apes also continued there, not just in Africa. Modern orangutans (the genus *Pongo*) in Borneo and Sumatra are the living descendants of those ancient apes (Figure 22).

The evidence from dated fossils on the one hand, and the pattern of changes in the DNA and proteins of living species on the other, complement one another. Where the relations of fossils of ancestors of living animals are obscure, molecular genetics can sometimes come to the rescue. While there is never a direct timestamp on genetic data, dated fossils can provide estimates of rates of genetic change. It turns out that the rate of DNA change on different branches of an evolutionary tree is similar. This fact, along with dated fossils on a particular branch, allows one to use the number of DNA differences between different species as a kind of molecular clock. The clock is regular enough to permit estimates of the time when different species' evolutionary ancestors first diverged and began following their separate evolutionary pathways.

Molecular-clock estimates indicate that the branch that eventually gave rise to the modern orangutans separated (probably somewhere in southeast Asia, according to fossil evidence) from the one that gave rise to the modern African great apes (gorillas and chimpanzees and humans) about 15 million years ago. In Africa, the gorilla branch separated from the chimpanzee + human branch about eight million years ago. The

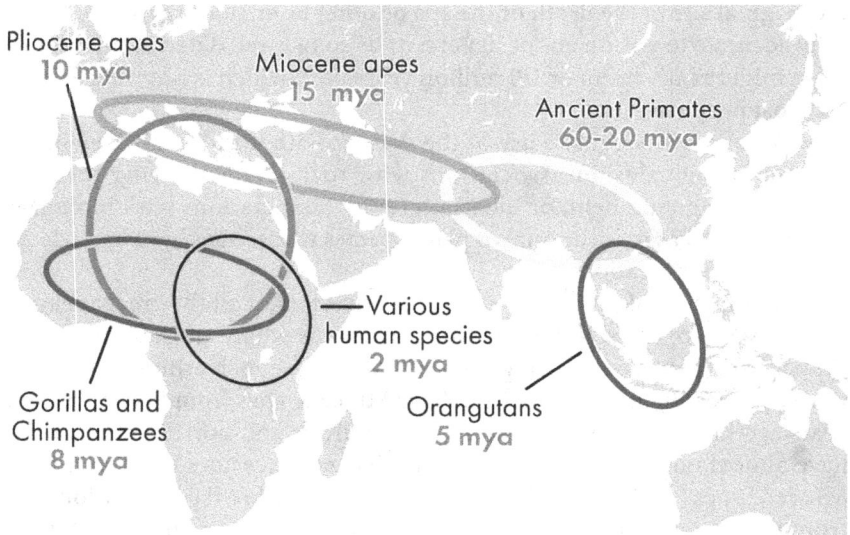

Figure 22. Evolution of the great apes according to approximate times of origin and geographic locations of fossils and living species, shown as a topographic map: lighter boundaries represent "lower" (earlier) stages of evolution; darker boundaries represent "higher" (later) stages of evolution (mya, millions of years ago).

chimpanzee and human branches separated into two evolutionary pathways about seven million years ago, one leading to modern chimpanzees and the other, eventually, to modern humans. This history makes the chimpanzees our closest evolutionary cousins.

The Genus Homo *and Its Ancestors*

The oldest fossils of the species *Homo sapiens*, parts of old skeletons very much like those of modern humans, are all from Africa. They are mostly from east Africa, although there are some from northwest Africa and some from south Africa. Ancient *Homo sapiens* may therefore have been widespread throughout the continent. The earliest fossils of our species are a little over 300,000 years old, which we can provisionally take as the approximate age of our species, unless even older *Homo sapiens* fossils are found. If the whole course of all life on Earth up to now were a 26-mile marathon, we humans would have joined the race just 10 feet before the finish line, the present time.

Although fossils of *Homo sapiens* older than about 300,000 years have not been found, there are other, older human-like fossils that might have been our African forerunners. *Homo heidelbergensis* is at least 800,000 years old. *Homo ergaster* is almost two million years old. Fossils of *Homo habilis* dating to slightly older than two million years ago have been uncovered in east Africa. All these species, along with a few others, are similar enough anatomically to be considered various species in the genus *Homo*. All have a flattened face and smaller teeth than earlier apes, characteristically human feet and pelvis, elongated legs and other indications of being upright walkers, a somewhat enlarged braincase and, from their associated artifacts, all were capable of making and using stone tools. Further back in time were other ape-like primates, likely precursors of the genus *Homo*, who also walked on their hind legs—although, from their anatomy, perhaps not exactly with a typical modern human gait. If they were not our direct ancestors, they were at least close relatives of those ancestors. *Australopithecus afarensis* ("Lucy") lived from four million to three million years ago and may have combined special abilities to climb trees as well as walk on the ground. *Ardipithecus ramidus*, who lived four or five million years ago, had some of the skeletal features of an upright walker, but had feet that would have been able to grasp branches, suggesting that much of the time it still lived and travelled in trees. *Sahelanthropus tchadensis*, from central Africa, an even older species, lived six or seven million years ago. It was likely a close relative of the ancient apes that were the ancestors of both chimps and humans It was about as tall as a six- or seven-year-old modern human child. The shape and size of its cranium was intermediate between that of a modern human and a modern chimpanzee. It walked upright.

Figure 23 indicates where these species are likely to have been located on our branch of the hominid evolutionary tree. (This tree is a summary of our present knowledge; such trees are always provisional to some degree, pending the unearthing of more fossils and possible realignments of existing fossils based on new measurement and interpretations.) In this figure, old species are near the bottom, dead and buried. Higher up are more recent species, also extinct. Species still alive today are at the dashed line at the very top, rising up into their unknown futures.

Most of the branches in the evolutionary tree in Figure 23 represent species long extinct. Fossils have been found for some of these; a sampling is indicated by the small black dots. Many species must have come and gone without leaving any fossil traces, at least none that have been discovered so far. The bare branches have nevertheless been included in the diagram because that's the way evolution ordinarily proceeds, with many related species existing at the same time, but most of them going extinct after a million years or so. The actual existence of at least some of the bare branches in the diagram is likely eventually to be justified by the finding of real fossils.

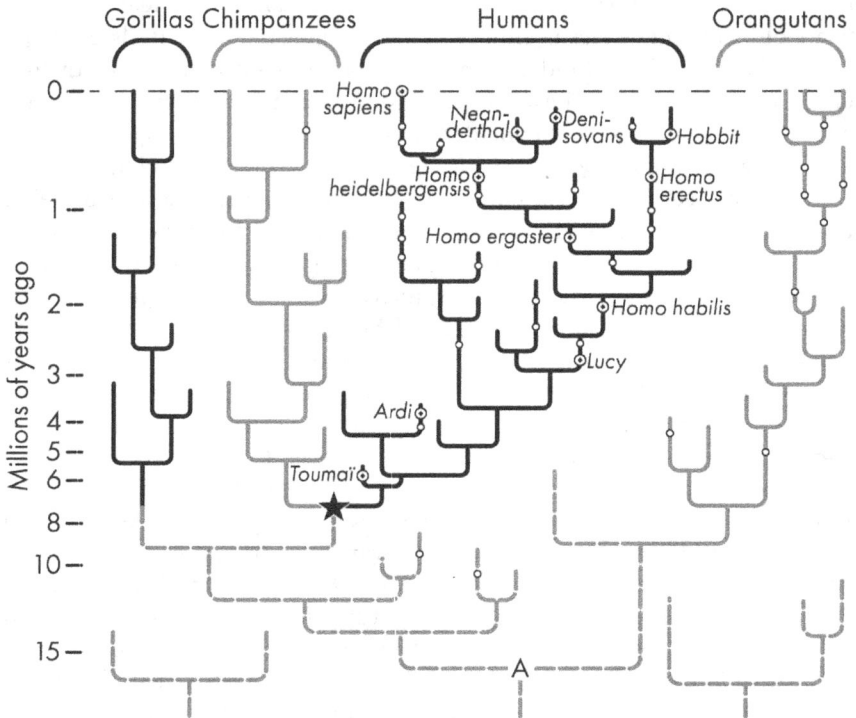

Figure 23. An evolutionary tree of the great apes. Gorillas, chimpanzees, and humans originated in Africa; humans spread from there all over the world. Orangutans originated in Southeast Asia and are still found only there. The letter "A" is the hypothetical common ancestor of all the great apes, which may have lived in southeast Asia. The star is the hypothetical common ancestor of chimpanzees and humans. The small black dots indicate some well-known fossil species. In addition to the standard species names given, there are "Neanderthal" (*Homo neanderthalensis*), the "Hobbit" (*Homo floresiensis*), "Lucy" (*Australopithecus afarensis*), "Ardi" (*Ardipithecus ramidus*), and "Toumaï" (*Sahelanthropus tchadensis*). See the text for further description and discussion.

A clever orangutan or gorilla or chimpanzee could trace its evolutionary journey from a common great ape ancestor (marked "A" at the bottom of the diagram) to its own species in today's world. The human reader could do the same and find his or her meandering evolutionary path from that ancestor to the present time.

Today there is just one species of *Homo*: it is us, *Homo sapiens*, now a worldwide species. Formerly there were also...

- *Homo neanderthalensis*, the Neanderthals;
- the Denisovans, so called from the Denisova cave in Siberia where a few bone fragments have been discovered (no official scientific name has yet been assigned to the species because there is so little in the way of fossil remains);
- *Homo heidelbergensis*, our probable immediate evolutionary ancestor who, according to the fossil evidence, lived from about 800,000 to 200,000 years ago;
- *Homo ergaster*, an earlier ancestor (its name means "working man," because of the stone tools found associated with its fossils);
- the tall *Homo erectus*, widespread in Africa, Europe, the Middle East, and Asia, extinct a mere 100,000 years ago;
- the short-statured *Homo floresiensis*, "the Hobbit," less than four feet tall, whose fossils were found in 2004 on the Indonesian island of Flores;
- the very primitive *Homo habilis*;

…and several other human species represented by unlabeled branches in Figure 23. It must have been an interesting time to live, between one and two million years ago, with several species of humans roaming around. The period around 50,000 to 60,000 years ago would have been even more interesting; that's when we *Homo sapiens* and *Homo neanderthalensis* came in contact with one another in Europe and Asia. We even mated with one another, to judge from the snippets of Neanderthal DNA that are still present in modern-day human populations.

A possible ancestor of ours, even older and not yet quite human, is "Lucy," *Australopithecus afarensis*. Lucy's 3.2-million-year-old bones were discovered in 1978 by Donald Johanson in east Africa; the nickname came from the Beatles' song "Lucy in the Sky with Diamonds." The find is remarkable for consisting of 40 percent of the skeleton. Even older was "Ardi," *Ardipithecus ramidus*, notable for demonstrating that an upright posture is compatible with having feet still good for grasping tree branches.

The star toward the lower left of the diagram in Figure 23 indicates the position of the ancestor shared by modern humans and modern chimpanzees. That ancestor ape is not known from a specific fossil, but its close relative was *Sahelanthropus tchadensis*, nicknamed "Toumaï." For the time being, Toumaï is a reasonable stand-in for the common ancestor of chimps and humans.

In the grand evolutionary tree of life, all the species living today are the millions of living twigs at the ends of branches that preceded them, and which were once end twigs themselves. We humans are just one twig among many. We shouldn't consider ourselves the pinnacle of evolution. We do have a number of unique traits, and some unique DNA sequences not found anywhere else in the world—but the same can be said for any other species on the planet—elephant, hummingbird, rattlesnake, the Puerto Rican coquí frog, or the sea urchin.

Sea Urchin Interview

All the microbes, plants, and animals of this world are products of evolutionary processes: environmental change, gene mutation, and natural selection with a generous dash of genetic drift thrown in. Each species alive must have had something going for it; it must have become well adapted for survival and reproduction, or it wouldn't be here, now, in the modern world. In that sense, no one species is "better" than another. They are just different in their own species-specific ways—the mighty oak no better than the lowly dandelion, the butterfly no better than the cockroach, the human no better than the sea urchin (Figure 25).

Humans' place in the world would look very different from the point of view of a sea urchin. We could picture the sea urchin's perspective being revealed by the following imaginary interview.

Figure 24. *Lytechinus variegatus*, a species of sea urchin (sex not apparent here). Sea urchins live in shallow coastal waters. They belong to the group of animals called the Echinodermata (the "spiny-skinned ones"), which also includes starfish and sea cucumbers (iStock.com/Damocean).

REPORTER: How do you see yourself—that is, your species—in the grand scheme of things?

SEA URCHIN: *Philosophically speaking, our point of view is not unlike yours, but you humans are biased in thinking that you stand at the pinnacle of evolution.*

REPORTER: Please explain.

SEA URCHIN: *You humans have convinced yourselves that from the beginning the whole evolutionary tree was aimed at producing you.*

REPORTER: You have to admit that we humans are quite special. We can write poetry, play the violin, build tall buildings, live in Antarctica, fly to the moon. Don't you think that makes us superior beings?

SEA URCHIN: *No. First you define superiority in terms of those traits you happen to have and the things you happen to be able to do, and then you call yourselves superior because you have those traits and do those things. Your anthropocentric bias comes from your long helpless childhood, when you thought your parents were circling around you like minor planets around the sun. Even as adults you proclaim your primacy in the biological world (you even call yourselves primates!), if not in the whole Universe. You've invented your religions to convince yourselves that the Universe cares about you above all other species.*

REPORTER: What makes you different, then?

SEA URCHIN: *We don't have childhoods during which our parents take care of us for years, so we have dispensed with the I-am-the-center-of-the-Universe idea. I came from a cloud of eggs and sperm broadcast into the sea by my parents, whoever they were, and I immediately became an independent, free-swimming larva, before metamorphosing into the beautiful creature you see before you. I was on my own from the beginning, not woven into a social fabric the way you are. We are not as you humans see us, poor remnants of past life, a side branch of ancient creatures who aspired to become human and failed, doomed forever to crawl on the sea floor. We are not some sort of primitive creature who belongs to the past. We also have reached the present time just as you have, and quite successfully too. Here, I've drawn you a proper evolutionary tree. [See Figure 25.]*

REPORTER: So I suppose you think we are just another species on the planet, as you are.

SEA URCHIN: *Yes, and let me point out that you came late to the party. Your hominid lineage goes back only a few million years, and the genus* Homo *even less. Your vaunted intelligence, your unparalleled ability to exploit and despoil the environment, and your territorial, warlike nature may yet drive you to extinction. You have yet to prove yourselves a long-lasting species. Ours is a more ancient and honorable line, 200 times older. We carry in our genes and in our bodies the wisdom of ancient seas.*

REPORTER: Are you saying that you sea urchins are better than us humans?

SEA URCHIN: *Ah, such a human, competitive way of putting it! I'm only saying we sea urchins are different. We may have our own echinodermatocentric biases, but they are based on our own biology, different from yours.*

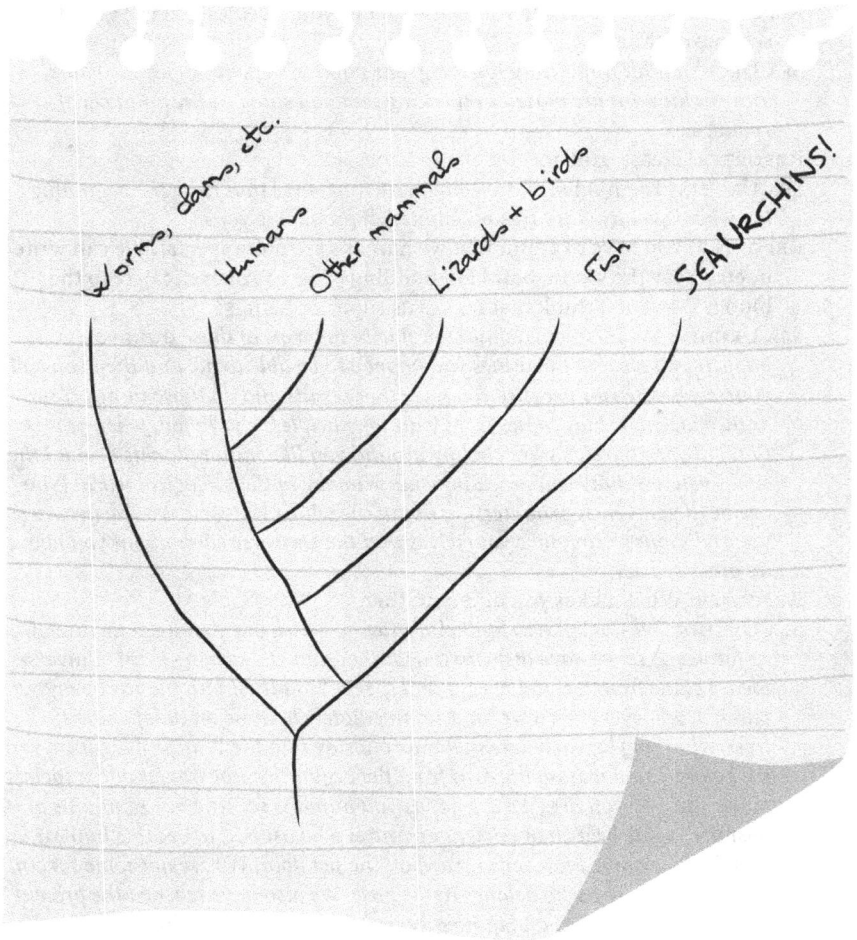

Figure 25. A sea urchin's sketch of an evolutionary tree, including the sea-urchin branch.

REPORTER: So any branch on the evolutionary tree could make a claim for some sort of superiority?

SEA URCHIN: *Exactly my point. If they are alive today, they are* all *at the top.*

REPORTER: Nevertheless, if we humans are the last and the most recent species to have evolved on Earth, isn't it reasonable to suppose that evolution was always aimed at producing us?

SEA URCHIN: *Just as you were not the first, so you are not the last, nor even the newest. Haven't you heard? There's a new species of butterfly that has appeared in the Colombian Andes just a century or two ago. Maybe evolution has always been aimed at* them!

REPORTER: Still, our intelligence, language, culture, and technology ought to count for something.

SEA URCHIN: *To be sure, those things are marvelous consequences of your biology. But you pay a high price for them. You have those awkward limbs sticking out, and you're always in danger of falling over. Most of your senses face in just one direction. That big melon head gives you all kinds of difficulties in childbirth. We sea urchins on the other hand (there's a human expression we have no use for!) have no Fallopian tubes, no uterus, no birth canal. We have an embryology that bestows on us the perfection of the circle and the sphere. Our nervous system is a ring; our senses are not concentrated in one place; we have no artificial "forward" and "backward"; we move equally well in any direction. Mouth at the bottom where the food is, anus at the top where the sea currents carry the wastes away. What could be a better arrangement? Is this not perfection? And you humans have only two gonads, while we have five. No bilateral symmetry for us! Radial symmetry is what we celebrate. And what is this business of having all your hard parts inside? All that succulent flesh on the outside, an invitation to any predator that happens by. How much more sensible to have a skeleton on the outside, and a spiny shell to protect our insides. Yes, you have your big head and intelligence and quickness of movement to protect you, but at what cost? In your case, as usual, Mother Nature had to make some tradeoffs.*

REPORTER: And in your case too, right?

SEA URCHIN: *Of course. We lack language, culture, and sports activities, but we don't bleed, we don't pump that sticky red fluid around in our veins. Our environment is as much inside us as outside. Less speed and energy required, soft and gentle movements only. That is our way. More in harmony with the pace of the Universe.*

REPORTER: Well, thank you for sharing your story. Our readers will be happy to hear your point of view and to know what you think.

SEA URCHIN: *Sharing, reading, happiness, point of view, knowledge, thinking—how alien all those things are! We sea urchins do not share, we do not read, we do not think. We do not seek knowledge, we* are *knowledge.*

6

Bigger Is Sometimes Better

Young Marcius: "A' shall not tread on me;
I'll run away till I am bigger, but then I'll fight."
—William Shakespeare, *Coriolanus* (1605–1608)

We're not in the class of the 100-foot-long 400,000-pound blue whale, nor the 12-foot-tall, 6000-pound African bush elephant. But we humans are still larger than the majority of animals around us—the croaking frog that keeps us awake at night, the bumblebee hovering over our flower bed, the hawk overhead, the sparrow searching for crumbs under the sidewalk table, the mouse in the kitchen at night, the neighbor's dog. It has been said that when all the animals on Earth are taken into account, the average animal is about the size of a housefly.

We are also among the largest of the primates, which include mouse lemurs and marmosets and bush babies (all of which you can hold in the palm of your hand), as well as many species of monkey that are about the size of an average cat or dog. Our size is our admission ticket to the group of primates called "the great apes."

Body size is a matter of the number of cell divisions during embryonic growth and the period of growth after being born. The fertilized egg I came from was about the same size as the one that developed into the mouse I saw in my kitchen one evening last month and the same size as the one from which came an African elephant. If you took a thin slice of adult liver or kidney and looked at it under a microscope, you could not tell whether you were looking at human, mouse, or elephant tissue; the cells are all the same size. An adult mouse has relatively few cells, a mere four billion. An adult human has about a thousand billion (10 trillion) cells. An adult African elephant has about a thousand trillion (one quadrillion) cells. It requires a longer time, cell division by cell division, to make a larger adult body (see Figure 26): only 10 weeks (including a three-week gestation period) to make an adult mouse, but about 18 years to make an adult human being or an adult elephant (elephant cells must divide faster than human cells). At 10 weeks the mouse has reached adulthood; at 10 weeks the human being is not even born yet. It is just getting started: new cells continue to be added for another thousand weeks. A body's reaching a certain size, small or large according to the species, depends on genes that regulate the number and rate of cell divisions, genes that are selected for the survival advantage they give the body in making its way in the world.

Being a large animal may be good in some circumstances but has its disadvantages. My body is a fertile breeding ground for bacteria, viruses, and other parasites. My skin's 20 square feet of surface area invites fungal infections. I would not be the first human being who from my natural standing height of nearly six feet fell off a curb and broke my collarbone. Barefoot in the grass or in my living room, I step on a sharp stone or piece of broken glass I hadn't seen because my eyes are so far from my feet. While

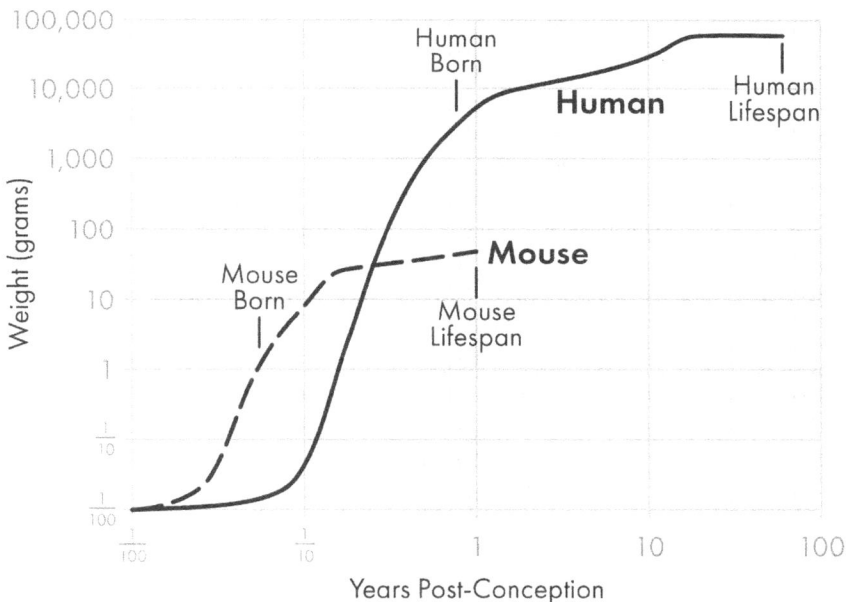

Figure 26. **The growth of the mouse and the human being, measured as weight in grams. The size of individual cells in both species is about the same, but the mouse has fewer cell divisions during its growth and ends up with fewer cells as an adult. The human has many more cell divisions during the period of growth and ends up with more than a thousand times as many cells. Logarithmic scales are used for both axes to accommodate two species of very different weights (graph constructed from data in** *Documenta Geigy, Scientific Tables,* **fifth edition, 1959, pp. 257ff., 267).**

in my prime I could jump from a standing position forward to a distance equal to my body length (professional athletes can do about twice that), I never could do much more than that trying to catch a bird or fish or escape the pounce of a tiger. But a jumping spider on the hunt can spring forward from a stationary position to about 50 times its body length. To catch food or escape from danger, flying squirrels can glide and bats and birds can fly. I'm not among these animals, alas. Given my human weight, the necessary wingspan and the forward speed to generate sufficient lift is beyond any reasonable expectation of what evolution could have accomplished. Angels might be up to it, but they don't have Earth's gravitational field to contend with.

During the long course of evolution most organisms have in fact stayed small. Given all the disadvantages, how did we, along with a few other species, get to be so large? The answer must be that there are compensating advantages.

In the beginning, almost four billion years ago, single cells were the only life forms on the planet. Even then size must have made a difference. If you happened to be larger than the other cells around you, you could eat them, and it's harder for them to eat you. If because you're larger you're able to move over longer distances, you have a wider hunting range and are not restricted to just what's available locally.

But single-celled organisms can't evolve to be bigger and bigger indefinitely; they can't follow the example of the giant Blob in the 1958 movie, which grew larger as it consumed everything in its path. There are practical limits to cell size. The Blob's increasing surface area wouldn't be able to keep up with its increasing mass; it would be unable to absorb enough oxygen and nutrients to maintain itself. Besides, once any cell's diameter exceeds a thousandth of an inch or so, the diffusion of molecules throughout the cell wouldn't be rapid enough for the chemical reactions in the cell to be maintained at any vigorous life-sustaining rate.

A few very large cells do exist. The yolk of a hen's egg is one large cell, but it's almost totally passive and grows only by being fed by the body of the hen. The same goes for a frog's egg (about the size of the "o" in the word "frog"), which does very little in the way of growth on its own but grows from a smaller cell only because the female frog pumps nutrients into it. Once fertilized by a sperm cell, the first act of any egg is to cut itself up into hundreds of small cells, which then take up the activity of growing, dividing, and differentiating to form the embryo.

One way to escape the limitations on cell size is to form cell aggregates. The first primitive colonies of cells may have arisen when dividing cells remained stuck together instead of going their separate ways. Those first cells may have had only a limited ability, individually, to perform different tasks. Or they may already have been able to carry out different functions in accordance with adaptation to the different environments the individual cells in the colony experienced. In either case, with the aggregation of cells came communication among them, and eventually some program of mutual inhibitions and division of labor. First in the course of evolution there probably was a simple differentiation of the reproductive cells from protective, nourishing cells. Then there came further diversification—the separation of cells that nourished from those that protected; of cells specialized in just sensing the environment from those that responded by moving the colony to a new place; of cells that provided pathways of communication from those integrated different activities. These were the primitive digestive systems, the simple skins, the modest sensory systems and muscle tissues, the elementary nervous systems and hormone-producing glands. All these developed because together they succeeded in more efficient production of the cells of propagation—the egg and sperm cells, the gametes.

We are the evolutionary descendants, from a billion years ago, of those first multicellular creatures. Our bodies and our brains evolved in the service of gamete production. The gametes free-ride without contributing anything at all to the body in which they reside in safety and security. The ovaries and testes do contribute to the adult body; from fetus to adult, they produce hormones that affect the form and function of the body. But the eggs and sperm themselves, no. It's as if the body, all unknowing, has locked its gametes away in a vault. Through its sliding door food can be passed; in the present they do nothing for us as individuals. But they carry the future.

In some situations, it is not larger but smaller size that can be advantageous. Where food is always scarce, for example, other things being equal, you might have more offspring if you can reach sexual maturity early. It is a general observation that when descendants of large continental animals migrate to the restricted land area of a nearby island, they tend to evolve smaller sizes. Not much larger than St. Bernard dogs, dwarf mammoths (a seeming contradiction in terms) existed on the island of Crete in the Mediterranean about a million years ago. There used to be dwarf elephants in Sicily and dwarf hippos in Madagascar. The smallest sea lions live in the Galapagos Islands. About 60,000 years ago there were dwarf humans on the island of Flores in Indonesia: *Homo floresiensis*, nicknamed "the hobbit," stood about three and a half feet tall. Because it takes a lot of time and a lot of food to grow large, old age and the increasing likelihood of accident may catch up with you before you can reach reproductive maturity. So if resources are limited or difficult of access, it is better to husband your resources and use them to produce more offspring rather than more bulk. But humans, one of the great ape species, didn't originate on an island but in the wide expanses of African forest and savanna where, if the means of getting it could be learned, there was plenty of food.

Across many branches of the tree of life, animals on continents appear to have conformed to the general rule that *bigger is better* and to have gradually increased in size over the millions of years of their evolution. The mammals that survived the massive meteor impact 66 million years ago were not much bigger than Chihuahuas, but from those diminutive creatures eventually evolved lions, grizzly bears, giraffes, hippopotamuses, elephants, rhinoceroses, and whales. Among the other small mammalian survivors were small monkey-like creatures, the ancestral primates. These were the ancestors of the modern great apes, including the ancestors of humans. Some of our ancient primate cousins remained small, retreating deeper into the African forests and taking up secretive nighttime foraging on vegetation and insects, just like their modern descendants, the lemurs (Figure 27). Other primates, distant ancestors of humans, took a different evolutionary path, moving out from the ancient forests of central and eastern Africa to the savannas, where food was more abundant and more accessible. This may have been one reason for the evolutionary increase in size. A large ape could range far afield to find enough grass, seeds, and roots

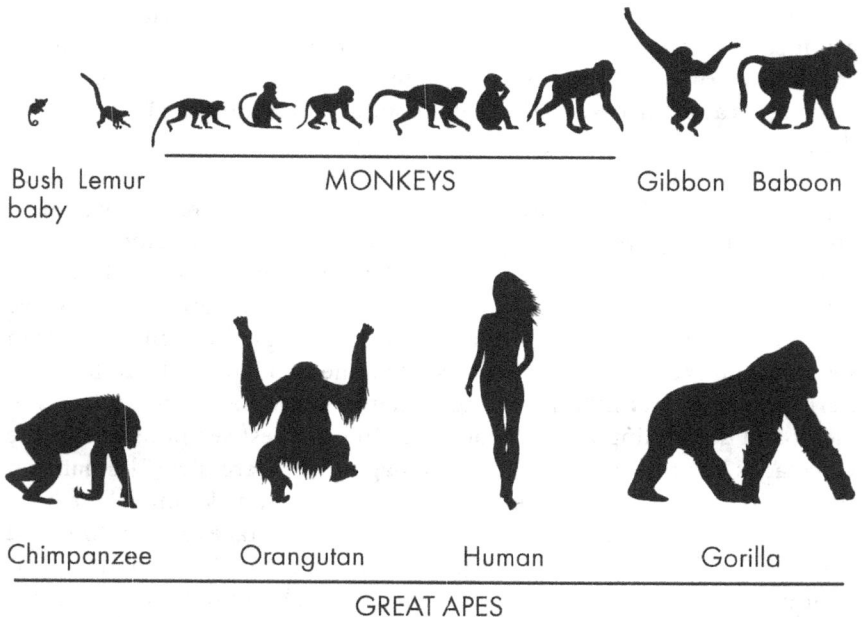

Bush baby   Lemur   MONKEYS   Gibbon   Baboon

Chimpanzee   Orangutan   Human   Gorilla

GREAT APES

Figure 27. Sizes of primates (iStock.com/Hein Nouwens and Niakris6/Shutterstock.com; human silhouette courtesy Milena Andonov).

to maintain itself, and grab the larger share for itself. A large ape could kill smaller animals and eat them. A large ape could fend off predators. A large ape could establish and defend its territory. A large male ape could be successful in fighting off his smaller competitors for the most beautiful and the most fertile females. Large apes might have been more conspicuous and more susceptible to attack by packs of predators, but they could make themselves safer by associating and cooperating with fellow apes, by evolving social behavior.

Because of a gradual change in climate that opened up the African terrain to sweeping plains and savannas, which became host to huge herds of large mammals, the newly emerging large omnivorous apes had access to meat on the hoof for the first time. The availability of a new and abundant food source allowed some apes to give up their leaf- and fruit-eating ways in the forest and over thousands of generations evolve into a large, cooperative, communicative, indefatigable, upright, fleet-of-foot, spear-throwing, meat-eating killer of large four-footed game animals. The descendants of the persistent forest dwellers, the chimpanzees and the gorillas, are still there in the jungle and the forested mountainsides. The descendants of those other ones, the apes that spread into the savannas, became *us*. These large apish and increasingly intelligent creatures, thrive they (we) certainly did.

7

Bilateral Symmetry

Tyger! Tyger! burning bright
In the forests of the night,
What immortal hand or eye
Could frame thy fearful symmetry?
—William Blake, *Songs of Experience* (1794)

Like 99 percent of all animals, human beings are bilaterally symmetrical (Figure 28). One side of the body is the approximate mirror image of the other: two ears, two eyes, two nostrils, two arms, two legs, two testicles. There is more mirror symmetry inside: right and left sets of upper and lower teeth; tonsils and salivary glands on each side; two lungs, two kidneys, two adrenal glands, two ovaries, two lobes of the thyroid gland, right and left halves of the brain. Even individual vertebrae have bilateral symmetry, providing attachments for the ribs on each side, and arranged so that nerves extend out from the spinal cord to each side of the body.

It might seem that Mother Nature has given us double body parts as a kind of insurance. If a kidney or an adrenal gland or a lung fails, we have second ones to keep us going, even if imperfectly. If one testicle or ovary is injured, the male or female can still participate in making children.

Going blind in one eye or deaf in one ear is not the calamity it might otherwise be, because we have a spare to see or hear with. In these cases, however, as in the case of having two hands, two parts working together give us an ability we might not otherwise have. "Man the tool-maker" could hardly exist without two hands. Depth perception and the sense of the world's three-dimensionality would be impaired without binocular vision. Localization of the source of a sound would be more difficult with only one ear. For more ancient reasons, which have to do with forward movement, evolution has made use of bilateral symmetry in the development of some additional basic skills.

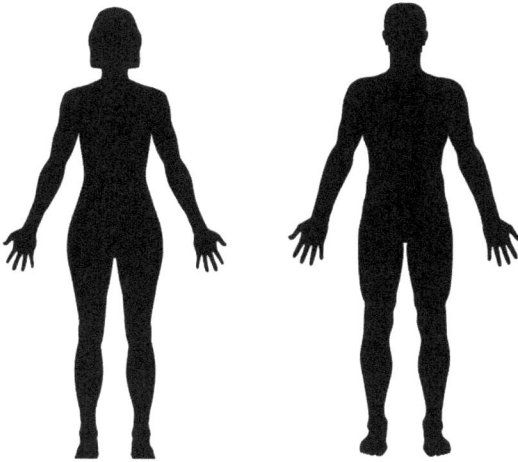

Figure 28. Human beings, bilaterally symmetrical.

Aside from the apparent usefulness of having a spare part, many important organs don't have backups, so there must be a different principle at work, not Nature's anticipatory wisdom. We have only one stomach, one liver, one gall bladder, one pancreas, one spleen. Having two of any of those organs would be useful, but we have only one of each; the reason is that these organs develop in association with the digestive system, which is a single tube running from one end the body to the other. Each of these organs arises embryologically as a mere side branch of that single tube, and so don't partake in the general two-sided symmetry of the body.

Even worse from the point of view of usefulness is having just one heart. I know several older people who would be pleased to have a back-up heart. Humans do in fact start out in early embryonic life with two "heart tubes," one on each side of the midline (Figure 29). Perversely, at the end of the third week of embryonic life, the two heart tubes come together side by side and fuse together to form a single structure, which then folds up to become the single heart we end up with. Mother Nature evidently doesn't really care for our long-term well-being, only for what works well enough for us to produce the next generation of our species.

If we had evolved on the Little Prince's tiny planet, B-612, with its very low gravity and thick atmosphere (an impossible combination), we might have taken the shape of hot air balloons, floating our way through life like dandelion seeds. Or if we had lived all our days suspended in mid-ocean in a watery, gravity-heavy planet, far from the surface above and the sea floor below, then there would have been no reason for bilateral symmetry ever to have developed. There would have been no up and no down, no forward

Blood flow

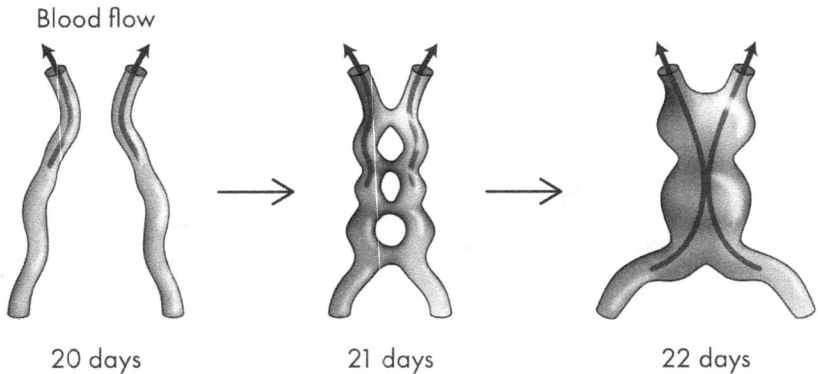

20 days   21 days   22 days

Figure 29. The embryonic heart starts out as a double structure.

and no backward, no right and no left. We would have evolved to be spheres or maybe cylinders.

But no, we're prisoners of our real evolutionary heritage, neither spherical nor cylindrical. We might not even ask how bilateral symmetry came about were it not for the animals who don't have it: sponges, sea anemones, jellyfish, and starfish (Figure 30). They are all water animals. Sponges, with their attachment to the bottom and their plant-like shapes, don't have much in the way of symmetry at all; like plants, they don't move around. Sea anemones and jellyfish have radial symmetry: parts of their anatomy are repeated many times around the circle. There's a top and bottom, but no front or back, no right or left side. Anemones are attached to the sea bottom. Jellyfish float midway between the bottom and the surface, and don't seem to care which direction they go.

Neither do starfish and sea urchins, bottom-feeding creatures, have much concern about what is ahead or behind. They have no head or tail; to them, any one of their five directions is the same. Interestingly, they start out their lives as bilaterally symmetrical larvae, and later undergo a drastic metamorphosis that turns them into these other-worldly creatures. Therefore, their symmetry is only five-sided secondarily. The change to their unique penta-radial symmetry is related to an evolutionary transition from an ancestral forward-swimming habit to a lifestyle of being settled down on the bottom. Any direction is "moving forward" for them.

What these asymmetrical or radially symmetrical creatures have in common is that they don't have to go looking for food in any particular direction. Sponges, sea anemones, and jellyfish wait for their food to drift by. Sea urchins and starfish slowly roam the sea floor, and the algal or molluscan food they seek is itself sedentary, and equally plentiful in all directions. They don't have to pursue it. Neither do they have to chase after

Figure 30. Some animals without bilateral symmetry: sponges, sea anemones, jellyfish, and starfish.

mates: stimulated by the number of daylight hours, the time of day, the abundance of food, or the romantic moonlight, they discharge their eggs and sperm into the sea water, trusting them to find one another.

Where there is no necessity for directional movement, there's no bilateral symmetry. In the early days of life on this planet, the ability to move forward went hand in hand with the evolutionary development of two-sided symmetry. Other kinds of body structure are possible for moving forward, although they are often less efficient, less maneuverable, and slower. Snails move forward at a snail's pace on a thin sheet of slime by means of muscular contractions of its single "foot." Some microscopic single-celled organisms move forward, as do sperm cells, by using a single locomotory organ located at the front or rear, like a single-oar scull or a Venetian gondola. An eight-person rowing boat can have its oars in alternate positions, right and left, an arrangement called "glide symmetry" or "glide reflection." A few ancient animals might have been built this way, but no known modern organisms are.

The earliest bilaterally symmetrical animals probably evolved on the floors of shallow seas more than half a billion years ago. They were not swimmers, but more like flat worms. They had an underside in contact with the seabed, and therefore also an upper side. Being animals, they hunted, foraged, or scavenged by crawling or slithering forward. They had a front end and a back end, as do all of their evolutionary descendants: fish, frogs, lizards, birds, rabbits, and all the rest. With a top, bottom, front, and back you automatically have a right and left side. For creatures above microscopic size, moving forward in a definite direction is most efficiently done with similar locomotory organs on each side, in something at least very

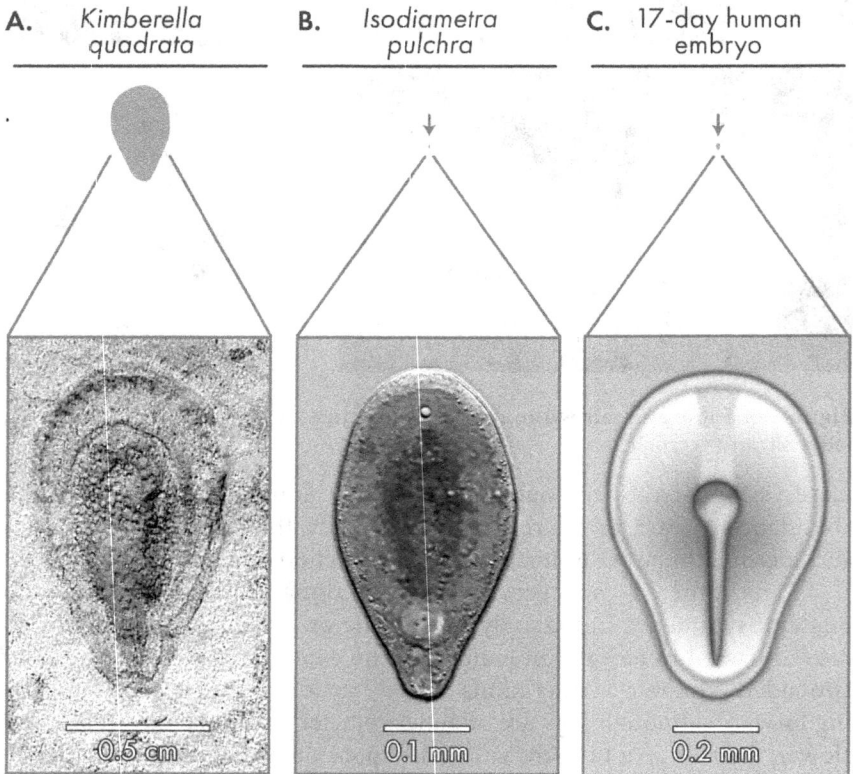

| A. Kimberella quadrata | B. Isodiametra pulchra | C. 17-day human embryo |

Figure 31. Primitive bilateral symmetry: A, *Kimberella*, from half a billion years ago; B, *Isodiametra*, a simple modern animal; C, drawing of a human being at the age of 17 days. In A, B, and C, the front end is toward the top of the image (photograph of *Kimberella quadrata* by Aleksey Nagovitsyn of the Arkhangelsk Regional Museum; photograph of *Isodiametra pulchra* from M. Chiodin, et al., "Mesodermal gene expression in the acoel *Isodiametra pulchra* indicates a low number of mesodermal cell types and the endomesodermal origin of the gonads," *PLoS ONE*, vol. 8, no. 2, 6 Feb. 2013, doi:10.1371/journal.pone.0055499; labels removed; human embryo, original drawing by Kate Baldwin).

close to mirror-image symmetry. Otherwise, like a rowboat with only one oar, any asymmetrical beast might go only in circles, hence the evolution of two symmetrical sides. For going somewhere, having movable lateral extensions of the body on either side is a good design.

Shown on the left in Figure 31 above is a fossil of *Kimberella*, a flat, worm-like, bottom-dwelling creature less than half an inch long. It lived more than half a billion years ago. We don't know anything about its internal anatomy. However, there is a similar little animal in the modern world. It is *Isodiametra* (Figure 31, middle). It lives among the sand grains on the floor of the north Atlantic Ocean. It is bilaterally symmetrical not only in its external shape but also in its internal anatomy: in its nervous system (with nerve bundles arranged symmetrically on either side of the midline), and in its gonads (with an ovary and a testis on each side). The 17-day-old human embryo (Figure 31, right) reflects the same ancient symmetry. It is a tiny, flat, oblong, slightly tapered disc. Some of its cells are beginning to migrate in opposite directions away from the midline, eventually to form a bilaterally symmetrical nervous system, gonads, and other internal structures. From the outside, at this stage, the little human looks like *Kimberella* and *Isodiametra*. Were it not such a transitory little organism and were it to be found living out in the world instead of being embedded in its little self-made space in the wall of the mother's uterus,

Figure 32. The Taj Mahal.

we might think it was just another simple creature like *Kimberella* and *Isodiametra*.

Evolution has taken the basic bilaterally symmetrical, flatworm body structure from half a billion years ago and elaborated its genes to erect the bilaterally symmetrical human being of today. In an analogous way, from its simple bilateral foundation the whole Taj Mahal was raised (Figure 32). Humans like bilateral symmetry in things they create.

So now when you read about a kidney transplant, or use your binocular vision to judge the position of the incoming baseball, or go rowing on the lake with your beloved (Figure 33), you might think of the primordial, bilaterally symmetrical, mollusc-like ancestor who first slithered through the mud on the bed of some ancient sea.

Figure 33. Bilateral symmetry in a rowboat outing (drawing by Margaret Evans Price, 1921, *Once Upon a Time: A Book of Old-Time Fairy Tales*, ed. Katharine Lee Bates, New York: Rand McNally; the Miriam and Ira D. Wallach Division of Art, Prints and Photographs: Picture Collection, the New York Public Library Digital Collections).

8

The Sexy Beast,
Part I

The Phoenix riddle hath more wit
By us, we two being one, are it.
So to one neutral things both sexes fit,
We die and rise the same....
—John Donne, *The Canonization* (1633)

The origin of the word *sex* is obscure. It may come from the Latin *secāre*, to cut or divide, referring to the division of a species into male and female individuals. How odd that in order to make one new individual, two individuals should be required! How much more economical it would be to reproduce without sex: I could forego any commerce with any opposite gender, I could forget about producing anything as senseless as an egg. I could clone myself, happily and prolifically producing offspring simply by budding off parts of myself, which would grow up to be more creatures like me (Figure 34).

The Biological Basics of Sex

There are a few organisms that reproduce asexually, by budding or other means. The bodies of the freshwater hydra sprout buds that grow into adults. Some simple worms split into two or more pieces and then regenerate the missing parts. The words "male" and "female" don't apply. A few species, among them some geckos and other lizards, are parthenogenetic, with egg-producing females only.

Most of the larger organisms, including humans, hawks, hookworms, and holly bushes reproduce sexually. Many single-celled organisms commonly have sex too, although there are no males or females, no sperm-producers or egg-producers, no embryos, just two similar cells fusing together. Sex must be a very old feature of biology, preceding even organisms made of many cells.

Other considerations aside, asexual reproduction is more efficient than sexual reproduction, by a factor of two. This can be shown by simple arithmetic. Imagine two goddesses, Aphrodite and Virgo, each of whom at the beginning of Time could produce a total of four offspring. Aphrodite and her descendants all have two female and two male offspring, and of course it's only the two females who produce the eggs that will carry the generations forward. By contrast, Virgo and her descendants dispense with males and produce four females at each generation—twice as many egg-producers. As one generation succeeds another, the expanding numbers of Aphrodite's and Virgo's descendants diverge, as shown in this table:

Figure 34. Asexual reproduction by budding (illustration by Makinze Jackson).

Total Number of Descendants at Each Generation

Generation	1	2	3	4	5	6
Aphrodite	4	8	16	32	64	128
Virgo	4	16	64	256	1024	4096

It should have been the descendants of Virgo, producing four daughters at each generation, who overwhelmingly populated the planet. The worldwide rule should have been reproduction without sex, by means of eggs only. Left in the evolutionary dust should have been sex-loving Aphrodite and her descendants, producing only two daughters at each generation. But, strangely, in the biological world sex is everywhere. There must therefore be some great advantage to allocating one's reproductive resources into making male offspring, when by simple arithmetic making only female offspring should be more productive. What's going on?

The reason sex is so widespread is not that it's the best method of reproduction—there are better ways to do that—but that it mixes genes from two different individuals. The evolutionary advantage is that, in the long run, the resulting genetic variety gives more offspring.

Why should this be? The answer is that the mixing of genes by means of sex is a way of counteracting the effects of mutation. The effects of gene mutation, that inexorably less-than-perfect duplication of DNA sequences, are generally neutral or occasionally even beneficial. But mutations often result in defective proteins and impaired functions of the cells of the body. As the generations pass, such mutations would gradually accumulate, leading to increasing numbers of defective offspring. Eventually the bloodline might die out were it not for the gene mixing provided by sex. The mixing of genes from two different people compensates for what would otherwise be an increasing number of bad mutations in the population. This is because, given their different ancestries, two different individuals are unlikely to be carrying exactly the same bad gene mutations. If genes from those two individuals are brought together in a single cell (by the act of fertilization), then whatever bad genes are present in the sperm cell might be compensated by the corresponding good genes in the egg cell, and vice versa. In addition, having a double set of genes from two different parents together in one child makes it possible for the child, when it matures, to assemble a better set of genes for the sperm cell or egg cell it uses to make the next generation's child.

Imagine, first, what might be called the Pryor–Wilder Principle of Defect Compensation. In *See No Evil, Hear No Evil*, Richard Pryor is blind and Gene Wilder is deaf. Individually flawed, together they make an effective team capable of foiling the plans of a criminal gang. The embryo, even carrying some defective genes, with a double set of chromosomes is able to make all the functional proteins it needs to develop normally, and for the cells to perform normally in the adult.

Imagine, next, something that is more than just *compensation* for defects: a reassembly of chromosome parts to make something for the future without the defects, in what might be called the Principle of the Moth-Eaten Tuxedos. Say my son needs a tuxedo for his wedding. I have available only two old tuxedos in my closet, but one has holes in the jacket and the other has holes in the trousers. Nevertheless, from the two tuxedos I'm able to assemble one usable one, and I can send my son off to prepare for the next generation of human beings. In the testis (or ovary), when sperm cells (or egg cells) are made, the chromosomes from the two parental sets pair off, and segments of two chromosomes of each pair are cut and spliced back together in new combinations. In this way, as a result of this special process in testis and ovary, chromosomes carrying fewer defective genes become available, as illustrated in Figure 35, A. These chromosomes can then be passed on, via the sperm or egg, to the new embryo.

A.

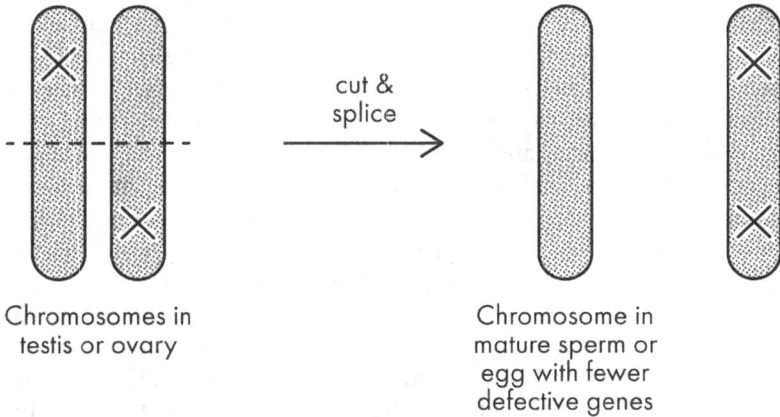

Chromosomes in
testis or ovary

cut &
splice

Chromosome in
mature sperm or
egg with fewer
defective genes

B.

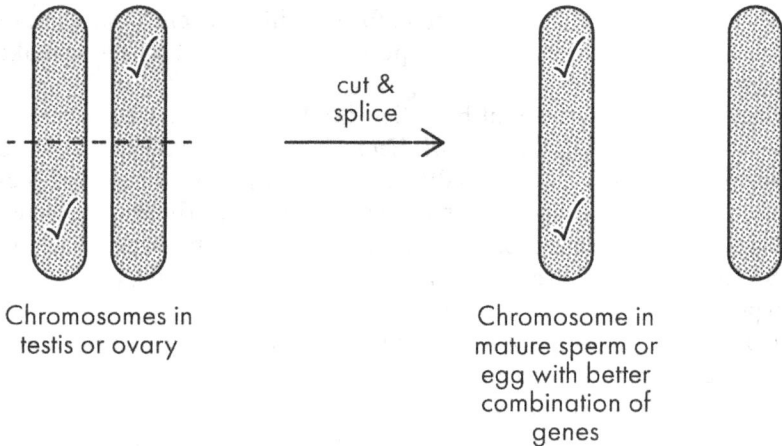

Chromosomes in
testis or ovary

cut &
splice

Chromosome in
mature sperm or
egg with better
combination of
genes

Figure 35. Making new chromosomes for the next generation when sperm cells and egg cells are formed in the parents' testes and ovaries.

The cutting-and-splicing process in testis and ovary, finally, can also give new chromosomes carrying better combinations of genes, which can be beneficial for the next round of embryos and adults, as illustrated in Figure 35, B, with what might be called the Principle of the Best Poker Hand. Besides individual gene mutations, this is another way that evolutionary changes can arise, potentially resulting in better adaptations for the next generation.

The bottom line is that if it's mere number of offspring that you're after, Virgo's method of reproducing asexually is better. But all her offspring are clones, genetically the same as Virgo herself and genetically the same as each other. Genetically uniform populations are vulnerable to contagious diseases and to other changes in the environment. Whole populations could be wiped out at a single stroke. With asexual reproduction, gene mutations keep accumulating in the descendants as generations pass, and the species eventually could become weighed down with bad genes. On an evolutionary timescale, Virgo's descendants, unlike those of her sexy sister Aphrodite, tend to give rise to short-lived lines. In the animal world asexually reproducing species do arise from time to time because of their reproductive advantage in the short term—some insects, some snails, some fish, a few salamanders and lizards—but on evolutionary timescales such species usually don't last very long. A sexually reproducing species with its gene mixing, on the other hand, contains great genetic variety, and therefore is generally less susceptible, as a species, to diseases and environmental fluctuations, and has a longer evolutionary life. We humans will stick with that method of reproduction; we don't have much choice, after all.

The mixing of genes could theoretically take place at any time in the life cycle, even after the two individuals had already developed into adults. In mythological times the amorous water-nymph Salmacis embraced the reluctant young Hermaphroditus, and their two bodies merged. Real gene mixing would have to have involved trillions of cells pairing off and then merging one by one. It would have been easier if they had merged in the usual way, a single cell of his fusing with a single cell of hers, but Hermaphroditus was too shy for that.

The best time for mixing genes from two individuals is obviously at the *very beginning* of embryonic development, when the male's single cell can fuse with the female's single cell—the whole point of maleness and femaleness, the whole point of sperm and egg. Once a new embryo is formed, every one of its new cells will have exactly the same mixture of the two parents' genes. Gene mixing and embryonic development are now inextricably joined, which is why sex and reproduction go together. We are descendants of Aphrodite, not Virgo.

Contrary to a popular metaphor, in no way are men and women from different planets. Their evolutionary origins and destinies—their bodies, their psyches, their behavior—are forever entwined. Without women, men would not have evolved to be what they are; without men, women would not be what they are. In the remote past the environment to which men adapted included not only the cold winter and the hot summer, the forest and the cave, the wild beasts to hunt or avoid, but also women and their ways. The environment to which women adapted likewise included males and their behavior.

The evolutionary liaisons between male and female are more intimate than merely adaptations to one another. As expressed in an ancient Chinese philosophical and religious symbol (Figure 36), male and female lines are never separate. Together they prepare the sexual landscape of the future, a landscape made up of the bodies and the behavior of both sexes. Men produce not only sons but also daughters, the terrain their sons will plow and where the sons will sow their seed. Women produce not only daughters but also sons, the spore-scattering stimulus for their daughters to bring forth their sprouts. For a woman to make a uterus and other parts of her reproductive system, she uses genes provided her by her father as well as by her mother. In order for a man to make sperm ducts and other parts of his reproductive system, he uses genes provided him by his mother as well as by his father. The female and the male are the yin and yang of sexual reproduction. The shape of each determines the shape of the other; each contains part of the other; together they make up the whole of the human species.

Sperm and Egg Logic

The gametes, the egg cells and sperm cells, are a lifeline thrown out by one generation to the next, life's only hope of continuing on after the death of the adult bodies.

> And nothing 'gainst Time's scythe can make defence
> Save breed, to brave him when he takes thee hence.
> —William Shakespeare, Sonnet XII (1609)

Figure 36. The concept of yin and yang, female and male, and other universal dualities, represented by the taijitu symbol dating from Chinese antiquity (Vicons Design, GB; thenounproject.com).

The gametes come from specialized "primordial germ cells" in the early embryo. Like skin cells, gut-lining cells, and blood cells, gametes are highly specialized cells, thrown out from their place of origin, and destined for a short life. After being released from the ovary or testis, they find themselves in an unaccustomed place where they have no future. Most of them die, surviving for only two or three days. Besides being too specialized for long-term survival, they have only a single set of genes. The sperm cell has hardly any energy reserves, and it has none of the molecular machinery for synthesizing new proteins, or to grow and divide to make new cells. The mature egg cell has energy stores and everything else it needs for synthesizing new proteins, but it lacks an essential "division center," which cells need to organize their internal structure for cell division, so on its own it could never divide to make more cells. The sperm has that division center, but nothing else necessary for cell division. However, if in the upper reaches of a Fallopian tube a sperm happens to meet an egg, then the two otherwise doomed cells fuse together, each supplying what the other lacks. A new cell is formed. The new cell, the zygote, takes on new life.

If there is to be sex, why two cells of such different sizes? A number of single-celled organisms also have sex, but often there are just two equal-sized cells that fuse together (after which the new cell continues dividing to make more single cells). By contrast, organisms composed of many cells make two sex cells of very different sizes—very large cells (we call them *eggs*, and the creatures that make them we call *females*) and very small cells (we call them *sperm* or *spermatozoa*, and the creatures that make them we call *males*).

The reason for two different sizes is the embryo. Like owners outfitting a ship at the start of a sea voyage, the two parents give the next generation's embryo what it needs to become a large, complicated, many-celled organism. The new embryo is provided with an abundance of molecular machinery, regulatory molecules, and energy-transforming structures to start it off on its journey toward becoming self-supporting. It is true that two small sperm-like cells could fuse together to make a tiny zygote, but the zygote wouldn't have the necessary provisions; it couldn't undertake the complicated process of growing up. On the other hand, two large egg-like cells could fuse to form a large well-provisioned zygote were it not for their difficulty in moving around and finding one another, whether in the water of sea or pond or the fluid of a reproductive tract. The best compromise, the evolutionary solution, is to have one small, streamlined cell and one large immobile but well-supplied cell. The likelihood of cell fusion is optimized, and the resulting embryo is well provisioned. Therefore, in advance of fertilization, the male's cells strip themselves of unnecessary parts and add a tail for efficient swimming, and the female's cells store up large numbers of molecules useful for embryonic development.

Unless sperm cells are made in large numbers, the chances of a sperm cell's finding an egg are small. This is especially true for aquatic animals that spew their eggs and sperm into the surrounding water. It's also true for mammals, which use internal fertilization. In humans, one ovulation gives one egg, while one ejaculation gives over 200 million sperm cells, most of which still get lost in the woman's reproductive tract. The man throws as many sperm as he can into the general vicinity of the egg, hoping one sticks. Also, because males tend to be promiscuous, a male's sperm cells could sometimes be competing with those of other males for access to eggs, another reason to make a lot of them. The more he makes, the better his chances of having offspring: his sperm can swamp out those of his rivals. So males with genes that result in more cell divisions in the developing testis will be the ones with the best chance of fertilizing an egg and making more offspring. And so it went, as one generation succeeded another: males evolved increasing numbers of sperm cells, many more than strictly necessary for fertilizing an egg.

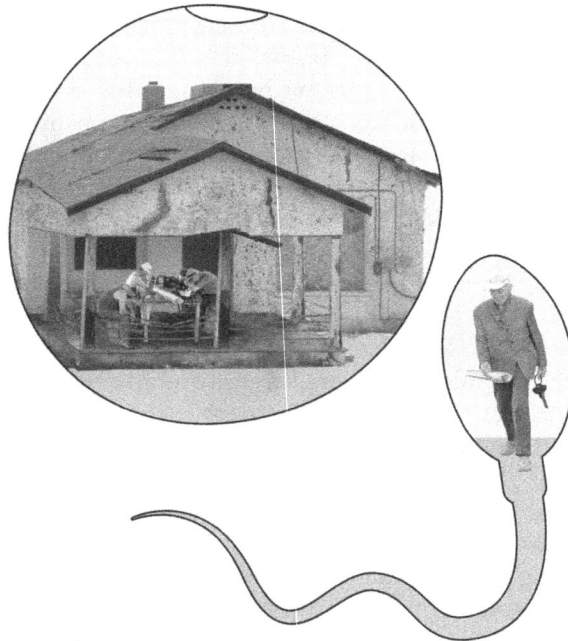

Figure 37. The egg and the sperm. The egg carries the materials and the tools and a single set of plans to make an embryo, but it needs the sperm with its keys as well as another set of plans to unlock the egg's potential. Neither has a future until it joins the other (photograph of building by Tony Santiago, with modifications made, Wikimedia Commons).

The egg cell just released from the ovary is like a house in the early stages of disuse, disrepair, and collapse (Figure 37). Inside, lying about in the gloom, are the workers, their tools and equipment idle, their plans for restoration locked up in a cabinet to which they don't have the key. The ejaculated sperm cell is a lively but homeless old man, likewise not long for this world. He carries with him some articles for which he has no personal use: keys to cabinets he has never seen, a set of tightly folded plans he has no use for. He stumbles into the house, and a transformation occurs. In short order the house becomes a going concern. Cabinets are unlocked, renovation begins. Copy machines are set running, the plans are distributed to all the new rooms as they are added. Where before there loomed only the prospect of death, the newly renovated building now radiates activity. In nine months, a small town is delivered to the world. In 15 or so years the sun is shining on a complete city with its communication systems, water and sewage departments, financial institutions, a police force, childcare centers, movie theaters and art museums and amusement parks—nerves and hormones, urinary and digestive tracts, mitochondria, an immune system, reproductive tract, a brain. The city is on the map.

The original house and the old man have long since disappeared, their identities absorbed into the new forms of growth and development.

The Mating Dance: The Logic of Fidelity and Promiscuity

Underlying the main features of our reproductive behavior are genes that have arisen during the course of evolution. Although humans are certainly influenced by societal norms, many details of our reproductive behavior are universal across all human cultures. In every culture women enhance their physical attractiveness in their dress and by personal adornments. In every culture men adopt symbols of power and exhibit their strength in physical contests. In every culture men's brains are wired to perceive certain female traits as attractive: prominent breasts, narrow waist, wide hips, a higher-pitched voice. Women's brains are wired to perceive certain male traits as attractive: broad shoulders, good muscular development, strong facial features, a deep voice.

The cells of men and women contain the same genes (except for the small number of genes in the male's Y chromosome which are concerned with the development of sperm cells). However, males from the fetal stage onwards use a specific group of genes to develop their male traits, and females use a different group in developing their female traits. There are slight gene-usage differences between the sexes even in parts of the body that don't seem to be directly concerned with sexual function, such as the brain. So it isn't only reproductive *anatomy* that is influenced by genes, but also reproductive *behavior.*

Males make advances, females accept or reject. This is a common story throughout the biological world. A male fish makes himself attractive during the breeding season by developing eye-catching coloration and by excavating a nest for the female; a female visits several males and their nests before choosing the one in which to lay her eggs. Male frogs congregate at pond's edge during the breeding season and begin calling; female frogs move in toward the pond and choose the larger males with the deeper voices, the ones more likely to sire many vigorous tadpoles and strong adult frogs. A male bowerbird, as part of his courting display, constructs an open-ended chamber of twigs and small branches and decorates it with brightly colored flowers, shells, berries and other objects. A female inspects the bowers of several males, assesses their quality, and then chooses a male and settles down for mating and nest-building. A female chimpanzee exercises a degree of choice by advertising her availability more openly in the presence of high-ranking males, thus improving her chances of getting high-quality sperm for her offspring.

In the human species the same principle is at work: *homo proponit, sed mulier disponit.* The underlying biology dictates that, by and large, men are always on the prowl for mating opportunities, even if not always overtly, and women are continuously maintaining a defensive lookout for unwanted sexual advances and deciding on who can and who cannot have access to her eggs.

On a Friday night in a singles bar, a man initiates some casual chit-chat with the woman on the next stool. A few minutes later he ventures, "So what's your name?" "Oh, no, you don't!" she says, "Next thing you'll be asking me for my phone number!" *Man proposes, but woman disposes.*

At the cotillion the young boys line up along one wall of the dance hall and the girls sit in chairs on the opposite side. A loudspeaker tells the boys to choose their partners. Many girls accept the proffered hands, but a few shake their heads, preferring to sit out the dance rather than dance with that particular boy.

The girl's choice to accept or refuse may be conditioned on the boy's behavior—he can approach with bravado or with politeness, and this may affect her response. She can refuse in a way to indicate, "No, I'll never dance with you," or in such a way as to say, "Not this time, but you might ask me again later." His further behavior may be determined in part by subtleties in the kind of response she gives. The interactions constitute a *mating system*, one in which the details of male and female behavior are intertwined.

This pattern of behavior—the indiscriminate pursuing male and the reluctant choosy female—has an evolutionary logic behind it. Fundamentally, the reproductive behavior of males and females is a consequence of there being so many more sperm than eggs in the world.

The human male, following biological rules established hundreds of millions of years ago, produces about trillion sperm cells over his lifetime. His reproductive output—the number of children—is limited by the total number of eggs he can manage to get fertilized, and not by the number of eggs one woman alone can produce. Genghis Khan (1162–1227) had one official wife but hundreds of minor wives with whom he is reputed to have fathered hundreds of children. In centuries past, potentates in China and India may have been just as prolific. A male's investment in his offspring, in contrast to the female's, is relatively small. He may protect and provision his family with his material resources, but once he has fought off his rivals, cornered his females, and donated a sperm cell to each egg, his work is potentially finished.

The biology of a woman is different. The number of offspring she can produce is limited by the number of eggs she can get fertilized. In her lifetime she is capable of releasing about 300 eggs, but her reproductive limit is really constrained by the human birth interval, about two years. Therefore, she could ordinarily have no more than about 15 children, assuming she could survive that many pregnancies. It is not enough for her simply to produce eggs and get them fertilized. Each developing embryo has to be protected and nourished inside her body. While her fetus is growing and her baby is being suckled, the amount of food a woman takes in and the energy she expends is double what she would need for herself alone. After a birth, even more is required of her if the child is to survive: months of nursing and years of protection and care. Until the child is able to fend for itself, the mother is a kind of double creature, or even a triple or quadruple one if she's caring for two or three offspring at the same time. Her investment in offspring is a heavy one.

Although there are major qualifications (to be discussed), the simple evolutionary gene picture is this (see Figure 38): genes affecting the male's behavior in such a way as to prompt him to cast his sperm cells as widely as possible are expected to result in a large number of his offspring spread throughout his territory. In the evolutionary past, the offspring would have inherited those genes, and his male descendants would have behaved promiscuously in the same way. These "male-promiscuity genes" eventually would have spread throughout the population until everyone had them—the males who actually used them, and the females who passed them on to their sons. This love-'em-and-leave-'em pattern of male behavior is widely seen throughout the animal world. A male stickleback fish constructs an underwater nest, entices a female to enter it, fertilizes her eggs, and guards the young when they hatch. Meanwhile he doesn't hesitate to lure other females into the nest and fertilize their eggs too. In Australia, small fairy wrens pair off and build a nest; eggs are laid, and then the male bird goes off in search of other females to inseminate. The male red fox establishes a den with a female, sires a litter with her, and patrols the boundaries of his territory looking for other females to mate with. On our own branch of the evolutionary tree, a small primate—the Alaotran gentle lemur of Madagascar—lives in small family groups consisting of one male, one or two females, and their offspring. In spite of what appear to be happy family groups, it seems that males not infrequently stray and father offspring in another group. And there is some genetic evidence that even gibbons, widely considered the most faithfully monogamous of primates, sometimes sire offspring with someone other than their mate.

The widespread biological imperative—that is, a behavior pattern established by evolutionary processes—is for the male to have as many offspring as possible by way of several females.

In Figure 38 (A, B), genes promoting male promiscuity (A) would at first sight result in more offspring than genes promoting male fidelity (B) and would be expected eventually to spread throughout the population. But there is this qualification: it's not just a question of the male's obtaining the maximum number of fertilizations, but the maximum number of surviving, vigorous offspring. It makes a difference how and where the male spends his time. Given that a male is a swimming, flying, or walking container of sperm cells, what kind of behavior on his part would lead to his having a larger number of healthy offspring?—leaving to find other females to inseminate, or staying home to help the mother defend and raise the offspring he has already fathered? The genes underlying one kind of behavior over the other are the genes that will spread throughout the population as the generations pass.

Because of the long-term helplessness of human children and all the care and feeding they need, the many females whom the male had impregnated might not always have been able to protect and feed her offspring adequately. Therefore, some of the children sired by the roving male would not have survived to adulthood. Genes promoting unbridled male promiscuity, "male-promiscuity genes," might in fact have found themselves in *fewer* surviving offspring and might therefore have *decreased* in number as the generations succeeded one another. Male behavior might have been modified by "male-fidelity genes," which would induce the male to stay with the pregnant female and the new mother and see that his offspring survived.

A paternal instinct therefore comes into play, weak as it might be compared to the maternal instinct. It is directed both toward the man's children

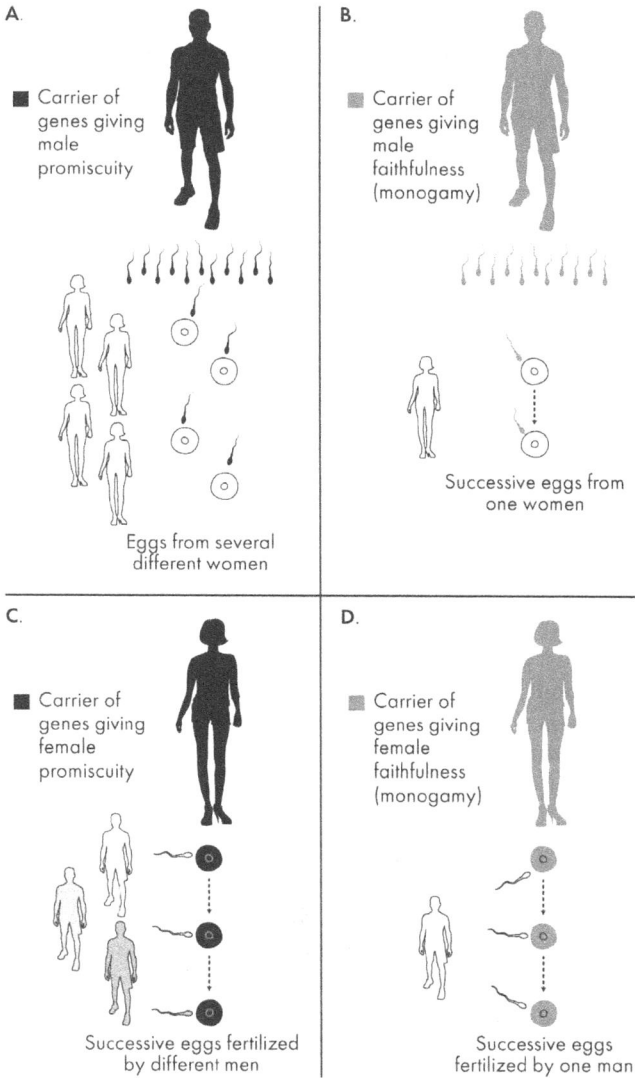

A. Carrier of genes giving male promiscuity

Eggs from several different women

B. Carrier of genes giving male faithfulness (monogamy)

Successive eggs from one women

C. Carrier of genes giving female promiscuity

Successive eggs fertilized by different men

D. Carrier of genes giving female faithfulness (monogamy)

Successive eggs fertilized by one man

Figure 38. The evolution of mating habits. Assuming there are "genes for promiscuity" and "genes for faithfulness," which ones will be passed on to greater numbers of children? (A) "Male promiscuity genes" might be passed on to more children than "male faithfulness genes" if all children survived (B). (C) "Female promiscuity genes" would not necessarily be passed on to any more children than "female faithfulness genes" (D) and maybe to *fewer* children if the female chooses a more caring male. See text for full explanation (silhouettes courtesy Milena and Assen Andonov).

and toward their mother, the children's primary caretaker. At an unconscious level the human male finds himself caught in a vise of contending evolutionary instincts: on the one hand, scatter your sperm cells widely; on the other hand, stay and provide for your female and your children. Both instincts are operational, providing more than one possibility for behavior. How a man actually behaves then depends on additional factors—upbringing, influences of parents and teachers and other forms of socialization, present circumstances and opportunities, and the weighing of possible consequences.

The human female, unlike the male, cannot increase the number of her children by being sexually promiscuous (Figure 38, C, D). She is the bearer of relatively few eggs, and she is constrained by the time and energy she has to expend to make an embryo and fetus and to raise the child. So any newly arising "female-promiscuity genes" would not be expected to be passed on to a greater number of offspring than could be produced by stay-at-home females—fewer, in fact, because she may have to raise her kids without reliable help. What a woman might evolve to do, however, is to select from the available males the one who displays the most trustworthy signs of being a good provider and protector, the faithful one. In this way, genes giving her the ability to discriminate among males would be passed on to all her descendants. Such genes would accumulate in the population because she's been choosy about the males with whom she copulates, and the children benefit. Such genes might have the additional advantage of providing an environment favoring the evolution of genes for fidelity in the male. The only possible advantageous outcome for promiscuous behavior on her part would be the occasional child who might receive better genes from a better biological father, raised under the protection of her unsuspecting mate. According to genetic studies, on average about one family in 10 includes a child fathered by someone other than the man in the family.

A woman has her own unconscious instincts contending with one another. Her "female-preference genes" influence her brain and her behavior with respect to what she looks for in a suitable mate. Having chosen a man who has stayed with her and provided for her and her children, her tendency is not to seek alternative mates. Yet having become better acquainted with him and his faults she can still find herself instinctively attracted to a different man with what she perceives as better qualities.

From such humble beginnings—the division of reproductive functions into two kinds of individual, those producing a few large nutrient-laden gametes and those producing many small motile gametes—came, over the course of millions of years, all the subsequent gene modifications that eased the union of egg and sperm and made the whole reproductive process more efficient. New genes and new gene-interactions supported the development of transport pathways for egg and sperm, new housing and new nutritive organs for the embryo, and glandular tissues of various kinds. We now have ovaries, Fallopian tubes, the vagina, a uterus, a placenta, mammary glands, testes, vas deferens, penis, prostate gland, and seminal vesicles. We think of all these as the primary differences between Woman and Man, but they are really secondary to egg- and sperm-production. From the same humble beginnings came the human male's predilection for female beauty (possibly a proxy for female reproductive potential), his unceasing interest in the other sex, and his wandering eye, his sexual aggressiveness, his predilection for promiscuous behavior. From these same beginnings came also human female's preference for male power, prowess, and physique (a proxy for his ability to provide and protect), and her license to accept or reject the male's attentions. Very little of this percolates up into the consciousness of either sex, but it's nonetheless the basis for much of human behavior.

It's true that there are societies in which marriages are arranged beforehand, not allowing much choice for either the woman or the man. There are societies in which children are commonly raised by relatives or other groups within the community, so the woman is not always dependent on a man for help in raising her offspring. These alternative social and political bargains may have arisen only with the growth of civilization a few thousand years ago and may not reflect the more primitive arrangements that prevailed in the small bands of humans that existed at the dawn of humankind. Our more basic instincts are legacies from the emergence of mammals 200 million years ago. They are parts of a system in which men and women, as far as their deepest instincts are concerned, have little choice about how it operates.

9

Up the Mammalian Path

...a curious illustration of the blindness of preconceived opinion. These authors seem no more startled at a miraculous act of creation than at an ordinary birth.... Do they believe that at each supposed act of creation one individual or many were produced? Were all the numerous kinds of animals and plants created as eggs or seed, or as full grown? and in the case of mammals, were they created bearing the false marks of nourishment from the mother's womb?
—Charles Darwin, *On the Origin of Species* (1859)

These are some of the evolutionary adaptations that characterize mammals: a high metabolic rate, a diaphragm to help breathing, fur to keep warm, red blood cells that lose their nucleus just before they enter the bloodstream, and, oddly, three small bones in the middle ear. (These bones are the *malleus, incus,* and *stapes,* known colloquially as the hammer, anvil, and stirrup. They are evolutionary size reductions, two of them derivatives of ancient reptilian lower-jaw bones.) Most of these traits are not things of which you would be much aware. Even less, unless you're a mother, would you remember the more defining mammalian traits: having been in intimate touch with that nourishment-transfer organ, the placenta, and deriving your later nourishment from mother's milk, produced by mammary glands.

The seed is planted in the womb; it grows to a certain size, too big for the mother to hold it in any longer; then it pops out and begins nursing (the mother is nearby!) in order to grow even larger. All around us we see other animals doing the same—dogs, cats, farm animals, zoo animals and, in films, elephants, lions, whales, gorillas, all giving birth and nursing, more or less the way we humans do. What could be more natural? It seems as if it's always been that way, the mammalian way. But mammals have been around for no more than about a hundred million years. There was a prior couple of hundred million years when land animals helped their offspring into the next generation in a different way, mainly by laying large yolky eggs, with no post-hatching nursing at all—something like what modern birds and most modern reptiles do. It's from those ancient animals that the first mammals arose. What happened? Where did the placenta come from, that intra-uterine device for transferring nourishment from the mother to the developing embryo and fetus before the entry of the young mammal into the outside world? Where did the mammary glands come from?

(We humans obtain fetal nourishment by means of a placenta, live birth, and suckling of the newborn by means of mammary glands. Because these are all female functions, one could be forgiven for thinking that males stand outside the whole evolutionary process of becoming a mammal. That's not true. Ancestral males contributed fully to mammalian evolution—by providing the X chromosome that makes many of the embryos turn into females and by providing half the genes used to construct the female reproductive system. Males may not produce the eggs; they may not nourish the young directly. But they always participated, and continue to participate, in feathering the mammalian nest. Without males there would be no females, no eggs, and no wombs.)

Origins

The first living creatures, single cells and simple multicellular plants and animals, appeared in the sea. Then larger sea creatures appeared. Eventually there were fish. Vegetation began to cover the bare rocky land. Insects followed, evolving from small shrimp-like marine creatures. Vertebrates that had evolved from ancient fish adapted themselves to make use of the new land environment, at least part-time. Some of those invaders of the land were amphibians, living much of their adult life on land, but staying near water. They still laid their eggs in water, and in water they grew up. Some of their descendants are still with us, our modern frogs and salamanders. Other invaders of the land cut their ties with aquatic environments altogether. They managed to obtain enough water from their surroundings and their food to survive. These ancient animals were neither reptile nor mammal. Having evolved from animals that laid their eggs in the water (like modern fish and amphibians do), they probably gave their embryos their start in life by providing the eggs with a large store of nutrients in the form of egg yolk.

The descendants of these early land animals followed two separate evolutionary pathways. One includes the modern reptiles and birds, whose embryos continue to carry out their development outside the mother's body in large, yolky, shelled eggs. On the other pathway there evolved a different system, one in which early embryonic development took place inside the mother's body. Modern descendants of this second branch are the mammals. The mammalian embryo starts from a small yolkless egg (a human egg is about one-third the size of the period at the end of this sentence). It is nourished initially by the mother's own products of digestion, via her bloodstream through a placenta. Later it is nourished from the mother's mammary glands. Figure 39 summarizes the two modes of reproduction as we know them from modern animals.

It was all a matter of how the young grew, how they obtained their nutrition all the way from the fertilized-egg stage to adulthood. To judge from what we see around us today, it appears that

- in some animals the mother put a large fraction of her resources for her young into large yolk-filled eggs at the beginning (making the eggs, in the case of birds, the largest single cells on the planet), and then either let them fend for themselves immediately after hatching, or continued feeding the hatchlings for a while until they were independent, and
- alternatively, the mother used her own food intake to nourish the embryo more or less directly (with the aid of a placenta) from the very beginning, keeping the embryo in an internal space (the uterus, or womb). When the young became too large to continue

EMBRYONIC DEVELOPMENT

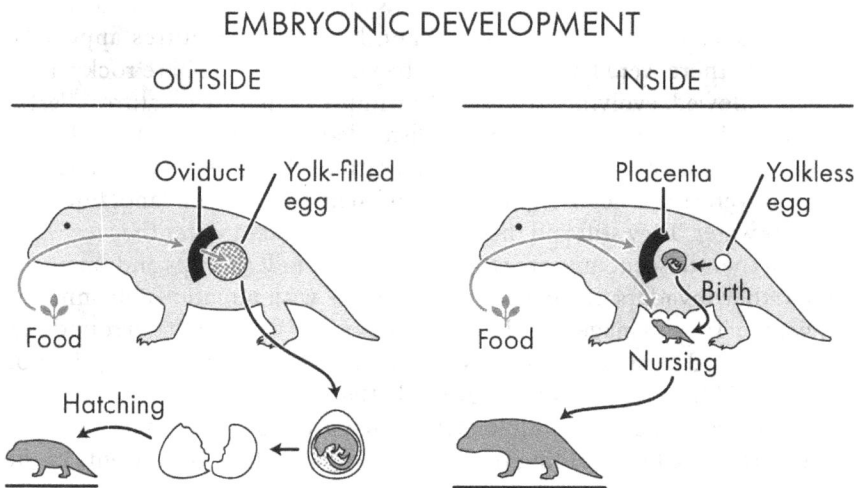

Figure 39. Two reproductive styles. There must have been an evolutionary transition from the one on the left, the more ancient one, to the one on the right, which is the mammalian (including the human) style.

growing inside, she expelled them and continued nourishing them with milk, sometimes supplemented with pieces of adult food, until they were mature enough to feed themselves.

To summarize (the mothers' main contributions are *underlined*; see also Figure 40):

- Grow OUTSIDE mother on *egg yolk she has provided* → (hatching) → independence
- Grow INSIDE mother on *her food* → (birth) → grow OUTSIDE on *mother's milk* → independence

The new mammalian lifestyle was first taken up by a few small animals that relied on a greater intake of food energy, a newly acquired ability to generate internal body heat, fur to help keep themselves warm whatever the outside temperature, a four-chambered heart for efficient pumping of oxygenated blood, a more continuously active lifestyle during the night as well as the day, and an enlarged brain capacity giving them the ability to learn from experience instead of relying solely on inborn instinct to get by. With such adaptations the mammals could adjust readily to local changes and to

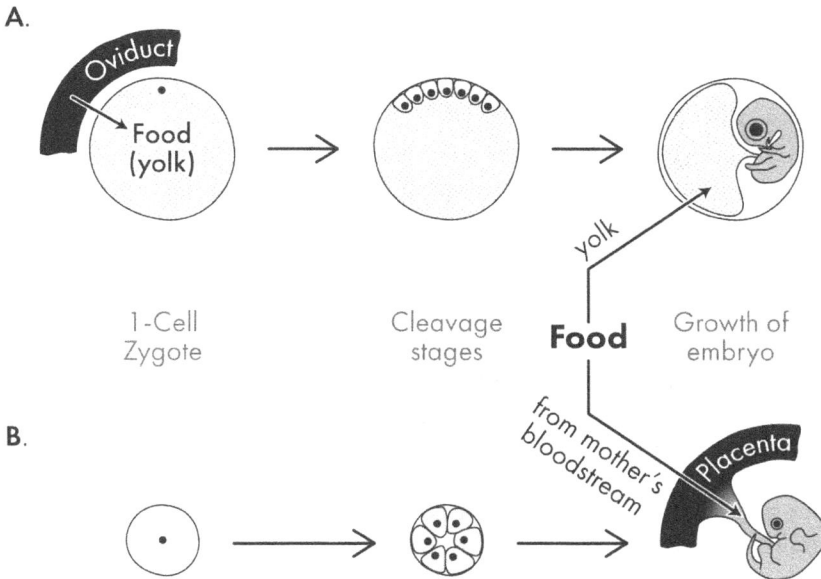

Figure 40. Early sources of nutrition for the growing embryo. Two methods are employed, one by reptiles and birds, probably the ancestral method (A); the other by mammals, more recently evolved (B). A black spot represents a cell nucleus.

varied environments. Mammalian mothers practiced a more intense care of the young, beginning by keeping the developing embryos inside her while nourishing them directly with her own food intake. After delivering to the outside world her young, too big to continue living inside but too small to take care of themselves, she continued feeding them by means of her mammary glands. For these new animals, these mammals, this meant expanding their range into new environments that were more or less closed off to reptiles, including the dark of night and the chillier crannies of northern and southern climes. The young were less numerous but more coddled and pampered, and so had a good rate of survival to adulthood.

These mammals might have remained small, timorous, and nocturnal, always underfoot of the larger dinosaurs, always the dinosaurs' dinner—were it not for the caprices of a restless Universe. Climate changes provoked by volcanic activity and a catastrophic asteroid strike 66 million years ago drove the dinosaurs and many other reptiles to extinction. This allowed the multiplication and diversification of the previously inconspicuous mammals. A large and diverse group of over five thousand mammalian species—rats, mice, bats, moles, rabbits, horses, whales, dogs, hyenas, bears, cats, lions, elephants, rhinoceroses, antelopes, and many others—currently are spread over the surface of the Earth. We live in the Age of Mammals. We are one of those mammals.

Mammary Glands and Milk

For an evolutionary biologist the transition period from a pre-mammalian egg-laying animal to a true live-bearing, milk-producing mammal would have been an interesting time. These "transitional" animals were in fact not just waiting to evolve into something better but were well adapted to their environments in their own right. They probably would have made eggs of a reduced size, with less yolk because the early nutrition of the embryo would have been provided directly from the mother's bloodstream. The pre-mammalian mother might have chosen moist soil or marshy vegetation in which to bury her eggs to keep them from drying out. She might have produced fluid secretions to coat her eggs with. After the young had hatched, she might have continued nourishing them with those secretions, modified to be more nutritious.

A version of this probable ancestral mode of reproduction can still be seen in the modern platypus of Australia and the echidnas ("spiny anteaters") of Australia and New Guinea, the only egg-laying mammals in today's world (Figure 41). The platypus, about the size of a housecat but with shorter legs, usually makes two eggs; they are not laid right after fertilization the way chicken eggs are but are retained in the mother's body in their leathery shells for about three weeks while the embryos are developing. After three weeks the eggs, each about the size of your thumbnail, are delivered to the world. The embryos continue developing and hatch about 10 days later. Then for another few months the young nourish themselves by lapping up milk the mother secretes into the fur on her underside.

Why platypuses, as well as the genetically related echidnas, have maintained such a means of reproduction for over a hundred million years is a bit of a mystery. Their ancient way of making and nourishing their embryos and their young must have been serving them well all that time, all the while environmental circumstances were driving other mammals to evolve the modern placenta. In order to counteract predation pressure, they might have been expected to have evolved a reproductive style that resulted in many offspring at a time, like squirrels with up to eight kits per litter, prairie dogs with up to 10, or rabbits with up to a dozen. But platypuses lay only two eggs at a time, and echidnas only one.

Maybe the combination of secretive egg-laying and burrowing behavior along with other anti-predator defenses—poison claws (platypus) and spines (echidnas)—just happened by chance to carve out a unique ecological niche that continued to serve them well, unchanged, for tens of millions of years. These animals, the modern monotremes, have hit upon an unlikely path of narrow solid ground through the treacherous quicksands of evolution.

If these strange animals are persisting with an ancient mode of reproduction that was in fact an intermediate stage in the evolution of placental mammals, then glands with nutritive secretions must have evolved first, and the true mammalian placenta evolved later, after the disappearance of eggshells and yolky eggs.

The human mother, after nourishing her embryo and fetus for nine months inside her womb, continues providing nutrition for her newborn

Figure 41. An echidna (left) and a platypus (right), egg-laying mammals. After the eggs hatch, the mothers nurse them with milk from their mammary glands (illustration by F. Mason, for *Guide to the Hall of Biology of Mammals*, Robert T. Hatt, American Museum of Natural History, 1901).

after parturition by means of lactation—the production of milk from her mammary glands, the female structures from which this whole class of animal is named. (The word *mammal* almost certainly comes from the young infant's opening and closing its lips in its primitive repetitious cry for the breast, *ma-ma-ma-ma*.... From that primeval cry come our words *mama*, *mom*, *mommy*, *mummy*, and *mammal*.)

All mammalian mothers, from mouse to moose, produce milk for their newborn. No matter what they eat when they grow up—rabbits munching grass, tigers tearing apart a deer, seals dining on sea bass, or humans sitting down to a dinner of steak and fries—mother's milk is the first meal of every mammal after its birth.

> The noble soul by age grows lustier,
> Her appetite and her digestion mend,
> We must not starve or pamper her
> With women's milk and pap unto the end.
> —John Donne, *To Sir Henry Goodyer*

Animals on the mammalian line gradually evolved to make smaller numbers of embryos, which were protected inside the mother's body instead of being vulnerable to nest-raiding predators. Shells were dispensed with. After a period of development, the fetus, prevented from further growth by space limitations inside the mother, was delivered directly to the outside world—small, shell-less, helpless and hungry. Unable to feed itself, it was given further nutrition by watery secretions from the mother's skin, secretions that may originally have been used by ancestral mothers simply to keep the eggs moist. The secretions may have contained salts, lipids, and natural antibiotics. Eventually, for the increasing benefit of the newborn, there were added proteins, lipids, and calcium. The secretions became milk. Lactation by the mother after the birth of her young is now the standard mammalian way.

Where did mammary glands come from? They are one of a number of ancient modifications of the vertebrate embryonic ectoderm ("outside skin") that include scales, feathers, hair follicles, nails, sebaceous glands, and sweat glands; they also include teeth and some of the salivary glands (embryologically derived from the oral cavity, which is formed by a sinking in of the ectoderm at an early stage). All these structures start out using similar gene networks to give a local cell proliferation and thickening of surface layers, followed by an invagination of the surface cells to form pits and eventually channels and ducts. In the glands, the deeper cells differentiate further to produce their specialized secretions, sweat, saliva or, in the case of mammary glands, milk.

In humans the mammary glands develop to the same immature stage in both the male and female fetus and remain the same in both sexes during childhood. At puberty, hormonal changes in the female lead to their full development, while in the male the mammary glands remain at a rudimentary stage. Why don't males develop fully functional mammary glands and share equally in the task of nourishing the newborn? It is an evolutionary division of labor, a question of efficiency in producing the most offspring. In a majority of our mammalian ancestors as in a majority of modern mammals, the male, having delivered himself of sperm cells, is long gone; he has nothing further to do with fathering his offspring. It's likely that any genes he's carrying that might impel him to take a greater share of the parental duties, including lactation, would be expected to disappear in the species' populations. Such genes would not be passed on efficiently: some of the female's offspring may not be carrying any of his genes at all, having been sired by some other sneaky male while he wasn't looking! Unless there are special circumstances requiring more active participation, males have greater reproductive success by leaving that female and siring more offspring with other females. Besides, the female, unlike the male, is always right there at the birth, immediately ready to continue caring for the newborn as it transits from inside to outside.

The first secretion that ancestral mothers provided their offspring may have arisen even before mammals evolved; it may have had the function simply of keeping the nest eggs moist. As mammals evolved and offspring were born live rather than hatching from eggshells, the secretions increased in nutritional value and came to be used to nourish the newborn, who could not forage efficiently by themselves. The new secretion—milk—would also have been easy to digest regardless of what their later diet turned out to be and would have been available to the young even during times when there might be a general scarcity of food.

Mother's milk contains vitamins, minerals, sugars, fats, and proteins, all important for the continuing growth of the newborn. The vitamins and minerals are transferred from the mother's bloodstream through the cells of the mammary gland to the milk. The mammary gland cells themselves synthesize sugars (especially lactose) as well as most of the fats and proteins and add them to the milk. The milk proteins, when digested by the newborn, are a source of amino acids for growth, and some of them also protect the infant from bacterial infections.

Among the more important milk proteins are the caseins; the word is derived from the classical Latin word for cheese, *cāseus*. When milk clots or curdles, aggregates of casein molecules separate as soft lumps or "curds" from the rest of the milk, leaving behind the liquid "whey" containing other proteins and other molecules. ("Little Miss Muffet / Sat on a tuffet, / Eating her curds and whey…")

To produce casein molecules, the lactating mammary gland's secretory cells use their casein genes. The particular mammalian forms of those genes are ones that no modern reptile has and that presumably no non-mammalian ancestor had. Where did these genes come from? We have a pretty good idea: evolutionary gene duplication, and a repurposing of the new gene copies in mammary glands.

Throughout the world, hard structures have calcium as an essential component. Calcium salts contribute to the hardness of clam shells, eggshells, coral skeletons, fish and reptile scales, teeth, and bones, as well as limestone, plaster of Paris, and cement. When the first vertebrates appeared a half-billion years ago, the cells of their scale-, bone-, and tooth-forming tissues secreted calcium salts into the external molecular matrix in which the cells were embedded. In ancestral biological systems, the cells of hard tissues would have synthesized and secreted into their local surroundings some sort of calcium-binding protein, a carrier for the calcium. Over the next few hundreds of millions of years, genes coding for proteins with calcium-binding properties continued to be passed down to piscine, amphibian, reptilian, avian and mammalian descendants.

With all the cell division and DNA replication going on all the time in the living world, genes here and there occasionally get accidentally duplicated. The extra gene copies become part of the genetic make-up of an organism and are passed on to subsequent generations. Such events have occurred in the evolutionary past of almost every species, and duplicate gene copies are widespread in both plants and animals. Sometimes, slow evolutionary modification and natural selection means that the extra gene copies may acquire new roles and may provide new advantages for the animals possessing them. The duplicated gene copies can be recruited to play new roles in new cell types in new anatomical places, even as they retain similarities with the ancestral forms by which their origins can still be recognized, together forming "gene families."

This is apparently what has happened over the course of vertebrate evolution. The casein genes of mammals, including humans, are members of a family of related genes whose proteins all bind calcium. As mammals evolved, some new copies of the original genes in this family were modified in such a way that their encoded proteins, even when they bound calcium, remained soluble in the liquid milky secretions of the mammary gland. These calcium-binding milk proteins, the new caseins, were not being used by the mammary gland cells themselves. They began serving as a way station for the storage and transfer of the calcium, which was tucked away inside aggregates of casein molecules. The milk secreted by the mother was taken in by the newborn, the casein molecules were digested in the newborn's stomach and intestines, and the calcium was released. In the growing infant it then went towards the making of new bones and teeth, its final and proper destination.

As you are reading this, milk is being lapped up and sucked into the mouths of young mammals everywhere—hamsters, hedgehogs, hyenas, hippopotamuses, and humans. In newborns' little digestive tracts the world over, casein proteins are being broken down, and the liberated calcium is being taken up by the cells of new growing bones and teeth.

Human milk is distinctive in one way that at first sight seems almost trivial, as well as somewhat puzzling: its overall concentration of protein is rather low, and it's slightly more dilute than the milk of the other great apes and the milk of many other mammals, including cows. That low protein concentration is correlated with a relatively slow growth rate of the suckling human infant compared with the newborn of many other mammals. This quality of human milk may be another evolutionary adaptation, one that appeared only on the human evolutionary line among the primates. Mammals in general, and primates in particular, are famous for their ability to learn from experience, as distinct from operating primarily or solely on instinct. Humans are especially good at this. A limitation in the amount of protein in human milk, especially whey proteins, may have been an evolved physiological mechanism that would have slowed down the infant's rate of growth, allowing the human brain to gain more experience from a longer infancy, and in this way provide a long-term advantage for survival and adaptation. The quality of human milk may thus have been a key element in human evolution.

The Placenta

In one of the earliest authoritative accounts of the New World, in 1511 Peter Martyr d'Anghiera described a belief held by the natives of Hispaniola, that when the spirits of the dead re-enter this world at night and masquerade as human beings, they can always be recognized by their lack of that telltale belly scar where the umbilical cord was attached. They have no navel, and it's because they never had a placenta or umbilical cord and were never human or even mammalian.

In reptiles, birds, and mammals including humans, the development of the embryo depends on extraembryonic membranes, one of which contributes to the formation of the placenta. The extraembryonic membranes are cellular sheets and partitions derived from small groups of early embryonic cells. They are not part of the developing embryo proper (which will grow into the adult body), but they remain connected to it through most of development. In mammals they support the growth of the embryo and fetus and are resorbed or discarded as the "afterbirth" at the time of delivery. The four extraembryonic membranes, each with its distinctive supporting function, are the yolk sac, the allantois, the amnion, and the chorion (Figure 42).

Early in human pregnancy the chorion merges with the wall of the uterus to form that special organ, the placenta. The main function of the placenta is to establish communication with mother, transferring nutrients and oxygen to the growing embryo from the mother's bloodstream, while waste products and carbon dioxide pass out of the embryo into the mother's bloodstream in the opposite direction.

A placenta has no hard parts that would be preserved in fossils, so the best we can do in trying to reconstruct a plausible scheme of its evolution, one that would take us from our distant non-mammalian ancestors to our present condition, is to compare different modern animals. That might at least tell us about the different kinds of placentas that exist out there in the world, and maybe some of them can give us an idea of where placentas could have come from.

In reptiles and birds, food and waste products are handled by the embryo's yolk sac and allantois, respectively. During development inside the eggshell, oxygen and carbon dioxide are exchanged through the shell (which is porous to gases) by way of two other membrane specializations, the amnion and the chorion. These membrane specializations disappear by the time of hatching. In mammals, however, the chorion has been evolutionarily modified. Whereas in reptiles and birds the chorion is plastered

REPTILE   MAMMAL

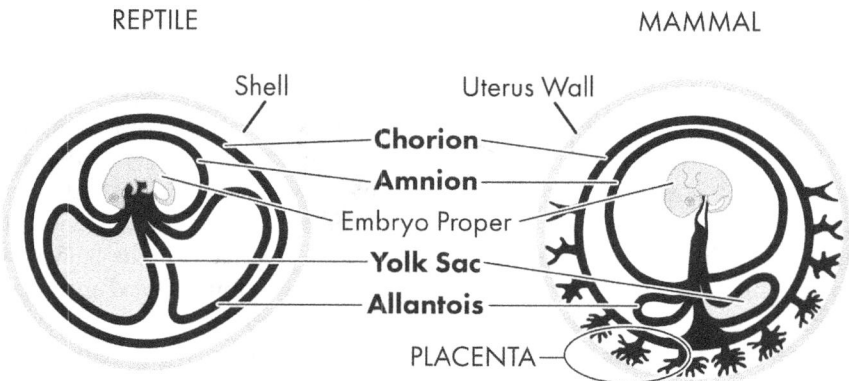

Figure 42. Comparison of the extraembryonic membranes of reptiles and mammals (dark structures). The amnion is similar in both. The yolk sac and allantois are reduced in mammals and play only a minor role in early development. The chorion is more fully developed in mammals, locally fusing with the inner wall of the uterus to form the placenta.

up against the inside surface of the eggshell, in mammals the chorion is fused to the uterus and partially merged with it, sending capillary-laden fronds into its wall in order to extract nutrition from the mother, thus forming the placenta (Figure 42).

The transition from a large, shelled egg to a small egg developing inside a uterus cannot have been that difficult, evolutionarily speaking. A few lizards, notably in the skink family, have taken on some aspects of the mammalian reproductive style. Like mammals, skink offspring develop from small shell-less eggs, are nourished by a placenta inside a uterus, and are delivered live to the outside world as small versions of the adult. (From that point on, unlike mammals, they have to forage on their own.) Most lizards first fill up eggs with a quantity of yolk and then deposit them outside to develop on their own. The circumstances that led to the skink kind of reproduction are not entirely clear.

Even though it required modifications of the extraembryonic membranes, female physiology, and uterine anatomy, the transition that set mammals on their path must have occurred a couple of hundred million years ago. It's not too hard to imagine the evolution of the mammalian mode of embryo nourishment from that of their ancient non-mammalian ancestors. It's safer, for the sake of the embryos, for mothers to retain their eggs inside their bodies for their embryonic development: a mother's instinct for self-preservation inevitably means embryo preservation as well. It's not hard to imagine a gradual thinning out and eventual loss of the eggshell, and a gradual transition to a smaller, yolkless egg, and an embryo that develops entirely within the body of the mother and is nourished there.

An ecological requirement for the evolution of this new mammalian mode of reproduction was that there be a fairly continuous supply of food available to the mother to nourish her developing and growing embryos. (By contrast, even in times of food scarcity, the reptilian mother can still make sure that each individual egg is well supplied with yolk; she has only to slow down her rate of egg production.) So seasonally adjusted mating and seasonal grazing or hunting behavior had to evolve along with the placenta.

10

Fur

In most cases, to be a mammal is to be hairy. Hair, or fur (the name usually used for shorter, denser hair that covers the body), is practically a defining trait for mammals; no non-mammal has hair. Having hair is part of the evolution of the ability to regulate body temperature, especially to keep warm at night and in cold climates. Hair allowed these newly evolving mammals to live in places that other animals couldn't and to stay active even at low temperatures.

One hundred and fifty million years ago when the first mammals appeared, they were probably not much bigger than mice; they evolved fur in order to keep warm. If an animal generates its own internal heat, being small means it is especially vulnerable to heat loss, because relative to its size it has a large surface area. Imagine an elephant generating heat from its internal metabolism, and at the same time it is losing heat from its skin. Now cut the elephant into 300,000 small pieces (the same total mass as 300,000 mice) and cover all the newly cut surfaces with new skin. Think how much more surface area there is now for losing heat. So to conserve heat the first small mammals evolved special skin cells for making fur.

There are some instructive exceptions. Whales, dolphins, sea cows (manatees), elephants, rhinoceroses, hippopotamuses, seals, walruses, the naked mole rats of Africa, and domestic pigs—all have lost most or all of the hair that their evolutionary ancestors possessed. The losses make sense in light of their various lifestyles. In some circumstances being hairy would be disadvantageous; in others, alternative modes of temperature regulation serve the purpose. Whales, dolphins, manatees, seals, walruses, and hippopotamuses live all or most of their lives in water and being hairy would be a disadvantage for them—although the sleek-bodied otters, fur seals, and platypuses manage quite well with their bodies being covered with short fur, which, like land animals, they may use to keep warm. Being without fur also makes sense for naked mole rats; they live their whole lives underground and have only a few hairs and bristles. Body hair would also be a problem in a hot environment, especially for large animals, for they have less surface area relative to their body mass to dissipate body heat. Rhinoceroses cool themselves by having a few sparsely distributed sweat glands and numerous small capillaries in the skin to dissipate heat; they also wallow in mud when they can find it. Elephants don't sweat much except between their toes; they are wallowers in mud when it's available. They also dissipate excess body heat by flapping their ears and fanning themselves, and by adjusting the distribution of blood in skin capillaries in different locations on the surface of the body.

Pigs were domesticated about 10,000 years ago from populations of fierce and hairy wild boars. On farms the pigs had a better life, without the extremes of temperature their boar ancestors had to suffer in the wild. When the weather turned cold, farm life itself was like a warm overcoat they could wrap themselves up in. Over thousands of years of domestic comfort, they gradually lost their hair.

Hair Today, Gone Tomorrow

We humans are the naked ape among the hairy ones. The other 500 living primates are all hairy. No doubt the evolutionary ancestor of modern humans, some ancient ape, was hairy too; it isn't likely that our ancestors were hairless aquatic animals living their lives in water like dolphins or whales or walruses. (To be fair, Elaine Morgan's book *The Aquatic Ape Hypothesis* does lay out a case for our evolutionary origin from apes that were adapted to water. If so, perhaps there hasn't been enough evolutionary time for us, having re-adapted to an inland existence, to regain our general furriness.)

Nor is it likely that our evolutionary ancestors lived underground like the naked mole rats of Africa. They may, however, have lived in caves, creating for themselves the comfortable living quarters of domestic pigs. Interestingly, the scientific name of our closest evolutionary cousin, the modern chimpanzee, is *Pan troglodytes*—which translates from the Greek as *the woodland god (Pan) who lives in holes or caves.* The scientific name comes from the first descriptions of chimpanzees in the late 1700s. It probably referred not to where they lived—chimps don't live in caves—but to the fierce behavior, uncouth appearance, and loud chattering calls reminiscent of certain wild people called the Troglodytae, who, according to stories that reached ancient Greece from Africa, supposedly lived in caves. It's entertaining to imagine that such stories are a faint echo of an ancient time when *Homo sapiens* was indeed a cave-dweller.

Somewhere along the line, we humans became less hairy than our hairy ancestors and less hairy than our modern great ape cousins (Figure 43). Why? Possibly we betray our hot African-plains origin when hairiness would have been a disadvantage to cooling ourselves by sweating from a large smooth skin surface. And maybe, once our *Homo sapiens* ancestors migrated northwards out of Africa, there was no longer any advantage to keeping warm by being hairy once we began to use animal skins to cover our nakedness.

A rare genetic condition known as congenital hypertrichosis (from Greek, "at birth, a condition of excessive hairiness"), extreme cases of which are sometimes colloquially called the "werewolf syndrome," shows that the potential for general body hairiness is still there, buried in our genes. At first sight it might appear that something different has arisen in these unfortunate people, almost a new kind of genetic infection. A more realistic interpretation of the condition, supported by molecular-genetic studies, is that the genes for hair-follicle development all over the body are still carried by all the skin cells of everyone. In the *Homo sapiens* line over time, starting maybe less than a million years ago, regulatory DNA sequences evolved that almost everywhere over the body's surface,

Figure 43. Comparison of two young great apes, human and orangutan (photograph at right by Kocurik, Wikimedia Commons).

suppressed the activation of the skin cells' hair-follicle genetic mechanisms. In a few people, rare genetic mutations in these regulatory sequences bring these hair-follicle genes back into action in skin cells in uncharacteristic places all over the body, and hypertrichosis results. The basic genes for general hair development are always there in everyone, a heritage from our pre-human ancestors.

Leaving aside ideas of a watery or underground stage in our evolution, there are better ways to explain the evolution of our present relative nakedness. Here are some ideas (all plausible, and none of them mutually exclusive); most biologists favor the last one in this list.

1. If you're a gregarious animal living in crowds, having a lot of body hair might have the disadvantage of providing cover for parasites, which could more easily transfer themselves from one individual to another. Chimpanzees spend at least an hour or so every day grooming each other, removing fleas and lice. A big ape that evolved to have less body hair might be less burdened with this time-consuming activity, might contract fewer diseases, and in the long run might therefore have more offspring.

2. A potential mate's having less hair might have made it more likely that he or she would be chosen for mating purposes. Fewer visible scars could have meant more victories in fights; fewer obvious insect bites and a smoother skin could have meant less susceptibility to parasitic diseases. If such a look was in fact correlated with better strength and better health, and made a potential mate more attractive, the result might have been a greater number of healthy offspring—who would therefore also carry the genes for less body hair.

3. An alternative to having your own coat of hair is to make one from the skin of another animal. You can use it both to shield yourself from the hot sun and to keep yourself warm in colder climates. Our distant ancestors couldn't just walk up and ask some passing antelopes; they'd have to kill them. Once our ancestors learned to use all the parts of the large animals they hunted, using bone splinters as needles and sinews and arteries as fibers to stitch together pieces of animal skins, their own coats of bodily hair became unnecessary. The genes whose use by skin cells had produced dense hair follicles gradually became restricted to just a few places on the body. The earliest evidence for clothing comes from a cave at the southern tip of Africa, where stitching awls made from animal bones have been found, dating to about 70,000 years ago. Clothing may have been made even earlier. Genetic studies suggest that the human body louse, which lays its eggs in clothing, evolved as a new subspecies of

louse about 100,000 years ago, so humans may already have begun making clothing then. This is about the time that *Homo sapiens* first migrated out of Africa into the colder climate of Europe.

4. You don't need a hairy skin to keep warm on those long winter nights in your cave if you know how to make a fire. Even if you used fire only intermittently—only in winter, or only for cooking, say— having your own fur might increase your chances of catching fire yourself, thereby decreasing your likelihood of reproducing. Your genetically less hairy brothers and sisters would then be the more likely ancestors of future generations. There is evidence that *Homo ergaster*, a probable ancestor of ours, might have been using fire more than a million years ago, even before *Homo sapiens* appeared on the scene. Our hairlessness might have been inherited from them. Some artists reconstructing *Homo ergaster* in his natural habitat make him relatively hairless, others hairier. We're not quite sure whether *Homo ergaster* was hairy all over or not.

5. Having a lot of body hair is a disadvantage if you're a large animal that lives in a hot climate in an open environment and runs after your prey, as our African ancestors undoubtedly did. For cooling, our ancestors took the perspiration pathway. They reduced their body hair and made room for more sweat glands. Profuse sweating is especially important if you're a hunter that puts out a lot of heat-producing effort chasing other animals for long distances on the open plains. We are—or were in our evolutionary past—that kind of hunter. Sweating a lot, our ancestors had to live near lakes and streams, where they had access to plenty of drinking water.

Why aren't other primates like us? Some may not have taken the evolutionary pathway to hairlessness because they are small, like macaques and lemurs, lose heat more readily, and use their fur to conserve heat. Others don't generate heat by chasing down prey animals; instead, like gorillas, they are vegetarians. Still others don't readily pick up parasites from other members of their species: like orangutans, they live a solitary life. Some, like chimpanzees, may have assessed the suitability of potential mates more by behavior and social status than by a smooth, hairless appearance. Only on one branch of the evolutionary tree did primates—our ancestors—learn how to stitch animal skins together, make fire, and take up the domestic life. The evolution of our hairlessness may have been overdetermined: several or even all of these factors may have operated together to drive the modern human being to its present naked state.

Perhaps on some other planet near another sun in another galaxy, similar circumstances have led to the evolution of another naked ape.

Running contrary to the general evolutionary trend toward hairlessness is the retention of hair in some places on our body surface. There may be specific advantages in survival and reproduction this gives us. The long, dense hair on the top of the head—one of the especially weird features of the human species in light of our general hairlessness—might shield the head against direct sunlight and keep the brain from overheating. Eyebrows could help keep sweat from running into our eyes and keep dust away; they are also used in communicating emotions. Eyelashes are a barrier to dust and insects. The hair in the armpits and the pubic region may have persisted during the course of human evolution because it served to diffuse sexual scents—perhaps, in the long run, resulting in more children.

Early in embryonic development, testosterone from the developing testis promotes the development of male anatomy. After birth, it stimulates an increase in bone mass, muscle mass, blood clotting, and level of aggression. With its high levels in the male, testosterone also stimulates the growth of hair on the face and elsewhere on the body. Therefore, males the world over tend to be not only bigger and stronger than females but also hairier. Hairiness becomes an external sign of a good testosterone level and its accompanying strength, stamina, hunting ability and aggression. In the right amounts, these traits helped the male in protecting and providing for his female and their offspring. If ancestral females tended to choose higher-testosterone males—that is, hairier ones—for mating purposes, and if males with higher testosterone levels and more hair tended to be the ones most successful in competing for females, then their offspring would have ended up being better protected and better provisioned. As one generation succeeded another, males would have passed on their genes for higher male levels of testosterone as well as genes linking testosterone to hair growth. That meant more and more offspring. In this way those genes, and the female preference for males who carried them, evolved together. From earlier human species we modern *Homo sapiens* have inherited the male-hairier-than-female trait.

On the flip side, the female's smaller size, higher voice, relative hairlessness and smoother skin may have tapped into the male's instinct to protect his immature offspring. Females having those feminine and child-like traits would have been better protected than other females. These traits would likewise have evolved together with the males' preference for them—and would have become, on the female side, the predominant traits of the female of the species.

11

A Warm-Blooded Animal

"I burn, I shiver," said Jinny, "out of this sun, into this shadow." ... Said Susan, "But I am not afraid of the heat or of the frozen winter."
—Virginia Woolf, *The Waves* (1931)

I'm what you'd call an endotherm; I generate my own heat internally, like all mammals. I'm also a homeotherm; I regulate my body temperature to keep it constant, within limits, almost wherever I find myself and whatever the weather. Therefore, there is harmony among the rates of all the chemical reactions that keep my body functioning.

Occasionally the mammalian thermoregulatory system is defeated by circumstance. A young anthropologist described her experience with hypothermia while traveling in the Peruvian Andes. We arose before dawn and caught a ride in the back of an open-air truck. "The temperature was well below freezing, and the wind was fierce. The road was unpaved, with rocks all over. I had taken a Dramamine pill, but I started vomiting before it could take effect. I kept vomiting on and off for the next few hours. Somewhere along the way I lost my gloves, because I had taken them off to hold the plastic bag I was vomiting into. Eventually I lost the bag too; I couldn't hold on to anything because I was shivering so much. I vaguely recall sinking into a miserable sort of nauseous, freezing stupor. When we finally reached our destination, my companion took my backpack along with his and helped me off the truck. I sort of remember just trudging along with my head down, watching his feet ahead of me and just trying to follow them. We ended up at an outdoor market and sat down on a bench near where an Aymara woman was selling hot drinks. They got some hot chocolate into me, and after half an hour I felt my brain gradually wake up again."

Similarly life-threatening is a rise in body temperature into the 40°C (107°F) range. It has its own set of physiological and mental symptoms. Twice I have experienced early symptoms of hyperthermia. On a tennis court in Miami in a hot mid-summer afternoon, everything around me suddenly became blindingly bright, the way your surroundings look after a visit to an ophthalmologist when your pupils have been dilated. The edges of the clouds in the sky above were flashing neon green and pink. My mental faculties remained sharp enough to recognize this pretty hallucination for what it was. I sat down on a shady bench, drank a cup of cold water, and poured another one over my head. In a few minutes my normal vision returned. At another time, as a student staying in a flea-ridden hotel on the Left Bank in Paris, I was overtaken in the middle of the night by a rapidly rising fever which held me in its coils for 24 hours of wild hallucinations. My immune system eventually took over and returned me to sanity. I had no means of taking my temperature, but it must have been high.

Otherwise, night and day, winter and summer, my body temperature stays within a degree of 37° Celsius (98.6° Fahrenheit). That my 100-million-year-old mammalian ancestors, small opossum-like creatures, could do this helps to explain mammals' long-term success. Over the course of evolution, they were able to migrate from hot savannahs to temperate forests to frozen tundra, and from cool nights into the hot noonday hours or vice versa, extending their range of activity into times and places from which their reptile-like forebears were forbidden.

Mechanisms

Thanks to my mammalian heritage, I'm pretty good at thermoregulation. My ability depends on eating more on a daily basis than does a similarly-sized cold-blooded animal—a large grouper or a Burmese python, say—and therefore I have a more active metabolism. I'm also more active physically than most reptiles and fish. My high metabolic rate and the contraction of my muscles generate internal heat as a by-product, so my body temperature is maintained at a higher level than that of my normal surroundings. In addition, for the sake of producing even more heat and keeping my body temperature up when it's required, the membrane-transport and energy-producing processes in my cells have an adjustable inefficiency built into them, which produces extra heat as a by-product. These mechanisms mean that I don't fall out of my tree when the temperature dips a bit; I don't have to seek out a flat rock and bask in the sun, lizard-like.

If the outside temperature rises beyond my body temperature, or if I generate excess heat internally by extreme physical exercise, then heat-sensitive nerve endings in my skin and elsewhere relay the information to the brain. The hypothalamus sends out hormonal and nerve signals, dilating skin capillaries so that excess heat is more readily carried away from the body's surface, and activating sweat glands so that the body surface can be cooled by evaporation.

If on the other hand the outside temperature is very low, several heat-saving or heat-generating devices come into play: the capillaries in my skin constrict, so there is less blood flowing at my surface, and I lose less body heat. My teeth chatter and I start to shiver. The small, rapid contractions of the muscles of my body and jaws don't produce useful movement (I'm not chewing, and I'm not going anywhere) but they generate extra heat. In addition, some of the molecular machinery involved in energy transformation and membrane processes in my cells switches over to "useless" mode, continuing to run at full speed but now acting only as little heat engines. The body, caught in a blizzard, disengages the clutch but continues to run its engine at full speed just to keep itself warm.

The chemical reactions in the cells of the body are sensitive to temperature because of changes in thermal energy and the frequency of molecular collisions and also because they are regulated by subtle and sensitive protein molecules, the enzymes. A temperature change of just a few degrees can alter how an enzyme molecule interacts with other molecules, and it can appreciably reduce or increase the rate of the biochemical reaction that the enzyme controls. In order for the body to continue functioning smoothly, given all its interlocking biochemical reactions that are dependent on the controlling enzymes, a change in the rate of one biochemical reaction might require a corresponding change in the rate of another biochemical reaction.

To maintain normal body temperature, cold-sensitive and heat-sensitive nerve cells send information to different specialized cells in the brain, stimulating the release of different neurotransmitters and hormones aimed at different end organs and muscles and capillaries. The mammalian thermoregulatory system is complicated. It can't have been built this way right from the start. The first developmental steps in setting it up were probably clumsy and imprecise, a baby physiology just learning to walk. Millions of years of evolution from non-mammalian ancestors have brought the system to its present state of refinement and complexity.

Evolution

What set the mammals along this evolutionary pathway so many millions of years ago? What's the point of conserving and producing all that heat to keep the body temperature higher than the cold outside? Or dissipating heat when the body is exercising or has a fever, or when it's just too hot outside? Why do mammals regulate the body's temperature so precisely?

Snakes, iguanas, geckoes, and most other modern reptiles do not regulate their body temperatures so closely. They track the ambient temperature and experience a wider range of body temperatures than mammals do. Reptiles have lived this way for over 300 million years and have had plenty of time to evolve biochemical reactions and their controlling genes and protein molecules so as to interact harmoniously over a range of internal body temperatures. Reptiles nevertheless do keep them all in step as much as possible by means of a crude kind of thermoregulation, but they achieve it behaviorally rather then physiologically. Rather than sweat or shiver, they burrow, huddle together, slide into the water, or move into or out of the sunlight.

Modern mammals are different, although evolved from ancient reptile-like ancestors. Long ago, when dinosaurs still ruled the Earth, a few small reptile-like animals experienced some random genetic changes that lasted because they helped save them from their larger predatory cousins. The new mammals survived by escaping into the forests and using the nighttime hours for their activities. Foraging during the cool of night required that they maintain high activity levels and higher body temperatures. Their behavior gradually evolved to allow them to use the hours of dawn and dusk and even dead of night to feed not just on leaves, nuts, and fruits, but also on insects and small cold-blooded lizards and snakes. The carnivorous part of their diet was the source of the extra energy that enabled them to dispense with the warmth of the sun and produce body heat by themselves. Their higher body temperature was more suited to nighttime activity. Body fur, one of the heat-conserving hallmarks of being a mammal, might also have arisen around this time. The regulation of a high and constant body temperature had another direct advantage for embryos. What environment—in contrast to one subject to the fluctuations in temperature that an egg laid outside may experience—could be better than a warm and cozy uterus?

Not straying from their ancestral nocturnal habits are today's bats, rats, mice, raccoons, coyotes, hyenas, tigers, as well as some of the smaller primates such as bush babies, tarsiers, pottos, and lorises. From these old nocturnal mammalian ancestors also arose many of the modern-day descendants that re-invaded the daytime hours: squirrels, chipmunks, mongooses, meerkats, cheetahs, camels, horses, zebras, buffalos, antelopes, baboons, and most monkeys, as well as our great-ape relatives, the gorillas, chimpanzees, and orangutans.

Advantageous as it might have been for ancient nocturnal mammals to expand their foraging and hunting activities into the daylight hours, the transition would have brought new difficulties. The warmer temperatures of daytime would have meant greater heat absorption from the environment, especially if there was a large difference between their own body temperature and the high outside temperature. The ability of small animals in particular to cool themselves would have been taxed. Relative to their mass, they have lots of surface area to absorb heat from the environment. The main method of cooling available to mammals, if they are not to curtail their daytime activities by retreating to a burrow or the dense shade of the forest, is perspiration. In order to cool themselves effectively, they would have had to spend most of their time drinking water and sweating instead of foraging for food. The difficulties of hot daytime living could also have been minimized by additional genetic mutations having the effect of elevating the body temperature even further, into the high 30s. Such animals would have found themselves at a relative advantage: with less difference between their body temperature and the average ambient temperature, there would have been a lower net inflow of heat from outside, and hence less need for profuse sweating. This is apparently what happened. We don't know exactly when. The slow evolutionary process must have occurred tens of millions of years ago. Over the millions of years of evolution, ancient mammals probably crossed the day-night boundary in one direction or the

Figure 44. The temperature-dependent harmony of the body's biochemical reactions is like the harmony of the instruments in a string quartet (The Negro String Quartet of the 1920s: Felix Weir, Marion Cumbo, Hall Johnson, and Arthur Boyd; Smithsonian National Museum of African American History and Culture, Gift of Dr. Eugene Thamon Simpson, Representative, Hall Johnson Estate).

other more than once. In any case, most modern mammals inherited a resting body temperature in the 32–40°C range, where we all live now.

The mammalian thermoregulatory mechanisms required a great deal of evolutionary fine tuning, and we humans are now sealed inside high-temperature bodies. Significant deviations from our normal 37°C (98.6°F) would upset our internal biochemistry and kill us. Over the millennia, as mammalian physiological machinery continued to run at the higher temperatures, evolution would have gradually brought about genetic adjustments in the enzymes and other proteins to make them all function optimally at our prevailing temperature of 37°C. We are now evolutionary prisoners of our own body temperature, locked into a temperature-sensitive system of biochemical reactions. Even relatively small deviations in our core body temperature, up to about 40°C (107°F) or down to about 30°C (86°F), from disease or circumstance, overcome our capacity for thermoregulation and become life threatening. Then the synchrony of the body's biochemical reactions is disrupted, some reactions speeding up, others slowing down. The result is a slowly unfolding chaos: heartbeat becomes irregular, kidney function collapses, drowsiness sets in, speech becomes difficult and confused, mental muddle reigns. Eventually there is loss of consciousness and, in the time-honored phrasing of medical textbooks, "death ensues."

My body is like a string quartet (Figure 44). The rise or fall of a few degrees changes the pitch of the violin relative to that of the cello and that of the cello relative to that of the bass. The resulting discord kills box office sales, and my orchestrated existence is over. But as long as I stay inside my temperature-controlled chamber, all the parts are played in beautiful harmony.

12

Upright Living

> *...we are a plant not of an earthly but of a heavenly growth.... And this we say truly; for the divine power suspended the head and root of us from that place where the generation of the soul first began, and thus made the whole body upright.*
>
> —Plato, *Dialogues* (4th century BCE)

> *But something else was needed, a finer being,*
> *More capable of mind, a sage, a ruler,*
> *So Man was born ....*
> *All other animals look downward; Man,*
> *Alone, erect, can raise his face toward Heaven.*
> —Ovid, *Metamorphoses* (8 AD)

Not long ago I spent a couple of hours in a coffee shop, reading and casually observing a concentration of strange bipedal creatures. What was most striking about these creatures, unlike the large congregations of wildebeests on the plains of Africa, was their verticality. They were long—that is, tall—for their size; their hind limbs were fully half the length of their whole body. You might expect that they would have been built with a lower center of gravity so that they wouldn't lose their balance and topple over. I've heard that in fact they do fall over from time to time, more frequently than other primates.

Could it be that Mother Nature, anticipating the great expansion of populations of these creatures, had long ago decided to conserve space by storing them upright? Or maybe it was simply a means of positioning our heads closer to Heaven, as the quotations above suggest. Or, more likely, bipedality was a trait that evolved in our non-human ancestors a few million years ago because of some relation it had to basic survival and reproduction.

George Bernard Shaw, playwright, essayist and social critic, and H.G. Wells, writer of novels and non-fiction works and a passionate popularizer of science, were the best of friends. As far as biology was concerned, however, they were at opposite philosophical poles. Shaw believed that our evolutionary advances were the result of a life force aimed at progress and improvement, expressed in us as our *will*, pulling us forward into the future. In 1928 Wells wrote to Shaw, "My warmest thanks for your friendly letter.... Your criticisms are very wise & valuable & also you are, as ever, quite wrong headed." Evolution doesn't work that way, he maintained. Human will is the *product* of evolution, not the cause of it. Humanity is forever being pushed from behind by evolutionary processes.

At the breakfast table, my wife sided with Shaw. "A couple of million years ago," she said, "when our ape-ancestors were hungry, they probably looked around and saw animals running away and decided that they could chase them down by running after them and wearing them out. They decided that this could be done more efficiently in the long run on two legs rather than four." Decision-making as an evolutionary force!

Biologists, however, agree with Wells, whose thinking followed that of Charles Darwin. The explanation, one that fits better with the universe's indifference to human desires, is that human anatomy is the result of mindless genetic adaptations to the environments experienced by our distant evolutionary ancestors. Conscious decision-making doesn't come into it. It is evolution by natural selection that has produced this remarkable biological feature of ours, unique among mammals: our two-footedness, our *bipedality*. (The word *bipedalism* is used interchangeably with *bipedality*. I prefer *bipedality*, because *bipedalism* sounds too much like a disease or a mental aberration.)

Forelimbs and Hind Limbs

In humans, the task of moving from one place to another has been taken over entirely by the hind limbs. Except for their contrapuntal swinging, the forelimbs have been pretty much liberated from locomotion. (The locomotory uselessness of the forelimbs of *Tyrannosaurus rex* explains their much-reduced size; see Figure 45.) The forelimbs of the creatures I was observing in the coffee shop were nevertheless quite long: at rest they hung down to mid-thigh. The forelimbs rarely rested, however. They were constantly being used to reach for something or carry something. With their five fingers at the ends, they are good devices for carrying handbags and cell phones and, in the old days, clubs and hand axes. The hands were continually fiddling with some plaything or other or arranging the long fur on the top of the head, scratching the nose, stifling a yawn or a sneeze. Hands leaf through books as if books and hands were made for each other.

They are used to solve Rubik's cube puzzles, deal cards, play the piano, and accompany speech with continual gesturing. A right forelimb is extended to that of another creature, hands are briefly clasped, a brief up-and-down movement of the forearm follows. Arms are folded across the chest, or their distal ends are placed on the hips with arms akimbo, maybe to convey an attitude, or perhaps just to keep the blood from pooling in their lower ends.

So here we are, these creatures in the coffee shop: everything that's being done—ordering coffee and pastry, handing over the cup of coffee and the pastry plate, giving and receiving money, picking up a newspaper, walking over to a table—could presumably be done, although perhaps less efficiently, by an ape that moves around mostly on four feet like a modern chimpanzee. But we are now doing it all upright, forced into our verticality because several million years ago our ancestors adapted to open woodlands and grassy savannahs after the thinning out of their forest habitats. We live in our 21st-century civilization in bodies millions of years old.

Except when we are sleeping or engaged in such activities as swimming or tobogganing, our modern upright stance is now obligatory. Our anatomy dictates how we spend most of our waking hours. Our upright stance, however, comes with some physical drawbacks not suffered by habitual quadrupeds: strained back muscles, spinal arthritis, degenerating vertebral discs and disc herniation, and intermittent and chronic back pains—one of the most persistent complaints of our species. We also suffer inguinal hernia and uterine prolapse from the weight of the internal organs pressing on the floor of the abdomen. Worst of all from an evolutionary

Figure 45. Two animals who evolved to walk upright, their forelimbs no longer used for locomotion.

point of view is a unique human trait, the narrowing of the pelvis, an adaptation that supports an upright stance. The female of our species has a narrower birth canal than any of our modern-day four-footed ape relatives—and, as we know from fossil skeletons, narrower than that of our evolutionary ancestors. The human birth process is sufficiently difficult that a third person is often needed to help bring the newborn safely into the world and dangerous enough that both mother and baby sometimes die during childbirth.

What could the advantages of an upright posture have been that they outweighed all these drawbacks? What on earth made our ancestors get up on two legs and live almost their entire waking lives that way? It can't have been only a question of standing up from time to time to have a look around. Many African monkeys do that, but as they range over the landscape looking for food, they are still quadrupedal. It can't have been a question of rearing up on your hind legs and beating your chest to intimidate your enemies. Gorillas do that, but they still travel on all fours. At a symposium a few years ago, an evolutionary biologist suggested that it must somehow have been a matter of "stand up or die." More likely it was a matter of "stand up and have more children." Getting around upright rather than travelling on all fours must have meant, in the long run, more offspring than before, because an upright stance and a bipedal gait, in spite of all the disadvantages, extended our ancestors' foraging range and allowed them to become carnivores as well as herbivores. Because meat is a more efficient source of energy than roots, leaves and fruit, the richer food and metabolic energy accumulated throughout adolescence and adulthood could, by means of evolutionary adjustments, gradually be channeled into their reproductive physiology and having more offspring.

Fossil Evidence

We *Homo sapiens* are not the first primates to have walked the Earth on two feet. Four or five million years ago, long before modern humans appeared, several other species of humans and human-like apes were employing some sort of upright two-footed gait. Even fossils of partial skeletal remains can show one or more of the following signs of bipedality:

- the bones of the foot, especially an aligned, forward-pointing big toe, but also the heel bone and the arch of the foot;
- ankle and knee joints;
- the connection of the femur to the pelvis (the hip joint) and the shape of the pelvis itself;
- the shape of the backbone and its vertebrae;

- modifications of the shoulder joint, forelimbs and hands indicating freedom from a primary role in ground locomotion;
- relative lengths and structure of the bones of the forelimbs and hindlimbs; and
- the position of the foramen magnum, the big hole in the bottom of the skull from which the spinal cord emerges.

The position of the foramen magnum in fossil skulls is a record of how apes moved around in the evolutionary past. It's toward the rear of the skull in animals that travel on four feet and whose body is more parallel to the ground, but more centered at the base of the skull in an upright walker (Figure 46).

From what can be deduced from fossil skulls and skeletons, the more-or-less upright postures of many of the ancient apes were not exactly like that of modern humans, adapted to walking or running on the ground, but to still spending much of their lives in trees, their feet adapted to grasping, their arms to swinging from branches. Over the course of evolutionary time, the later apes appear to have spent more and more time on the ground and an increasing amount of time running—maybe from predators, or after more animal prey as their diet changed from purely vegetarian.

First in the primates of ancient forests, then in *Sahelanthropus tchadensis*, *Ardipithecus ramidus*, and *Australopithecus afarensis*, then later in the *Homo erectus* of the African savannahs, and finally in *Homo sapiens*, the processes of evolution molded the different parts of the body

Modern chimpanzee	Sahel-anthropus	Australo-pithecus	Homo erectus	Modern Homo sapiens

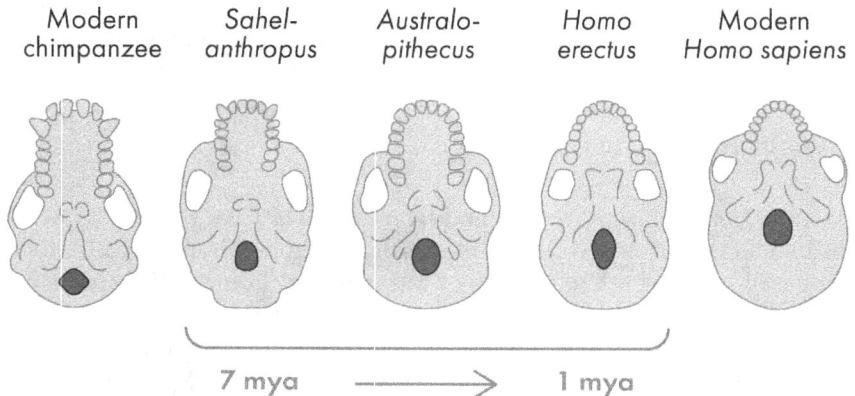

7 mya ——————→ 1 mya

Figure 46. Views of the underside of the skull of some modern and ancient apes, showing the position of the foramen magnum, where the spinal cord emerges. The foramen magnum is in a more basal position in apes that move around upright most of the time, more posterior in apes that travel on all fours, like the chimpanzee.

individually and gradually, small change by small change, but always in a coordinated manner that maintained the functioning of the body as a whole in the environment where it lived.

The anatomical features of these apes existed not as way-stations for the apes to come, but as the features of successful and well-adapted organisms in their own right, living for themselves, not struggling (and failing) to become human like us. These ancient apes were, like us, making their way tentatively across steppingstones of stability lying in the river of evolutionary oblivion. The individuals that made it from one stone to the next— those that successfully raised offspring and generated a successful line of descendants—contributed to the continuance of the species, giving it a future. There were no wild leaps into the unknown; each foothold in the environment had to be a solid place on which to stand. Without evolution's continual slow mutual adjustment of component parts, there would be only a piecemeal jumble of anatomy and physiology from which more offspring would be unlikely to be delivered—they and their descendants would fall into the stream and be washed away.

From the species that adapted to their local climate and their ecology with a degree of uprightness—originally in the service of surviving in greater numbers and having more offspring—eventually emerged, on our particular evolutionary branch, a separate evolutionary history of forelimb and hindlimb: different lengths, different shapes, different joints, different adaptations. The earlier four-limbed locomotory functions were partitioned into two and two. The forelimbs were free to evolve greater strength in grasping and throwing, and greater precision in manipulating objects. The hind limbs were free to evolve their flat, firm, springy, long-strided contact with the ground. Nowadays, unrelated to reproduction directly, there is cursive handwriting, embroidery, baseball, piano playing, the 400-meter hurdles, soccer, gymnastics, and ballet dancing.

The scheme below (Figure 47) shows a sampling of ancient hominins (species of humans and other ape-like species on our branch of the evolutionary tree), some of whom spent at least some of their time walking upright. From older to newer, they are "Toumaï," *Sahelanthropus tchadensis*; "Ardi," *Ardipithecus ramidus*; "Lucy," *Australopithecus afarensis*; and others without nicknames, abbreviated A. an, *Australopithecus anamensis* (possibly an evolutionary ancestor of Lucy); H. ha, *Homo habilis* (probably the first tool-maker); three species of *Paranthropus* (probable evolutionary descendants of Lucy, having gone extinct without leaving any descendants themselves); H. er, *Homo ergaster* ("working man" with more advanced stone tools); H. he, *Homo heidelbergensis* (likely our most immediate evolutionary ancestor); and H. ne, *Homo neanderthalensis* (the Neanderthals). An offshoot of the Neanderthals are the Denisovans (so-called from the

Denisova cave in Siberia where a fragment of a finger bone was found), not yet officially named *Homo denisova* (H. de) until more complete skeletal remains come to light.

To judge from the position of the foramen magnum in its skull (we don't have fossils of other parts of the skeleton), Toumaï probably walked upright much of the time, although maybe not exactly the way modern humans do. More skeletal parts are available for *Ardipithecus*. From them and other evidence we surmise that Ardi may have lived in open woodlands, spending as much time on the ground as in the trees. Its leg bones, pelvis, and backbone seem to have been structured for upright balance and support, and its flat foot adapted for walking but still capable of grasping branches, the big toe being at angle with the foot's long axis. *Ardipithecus*

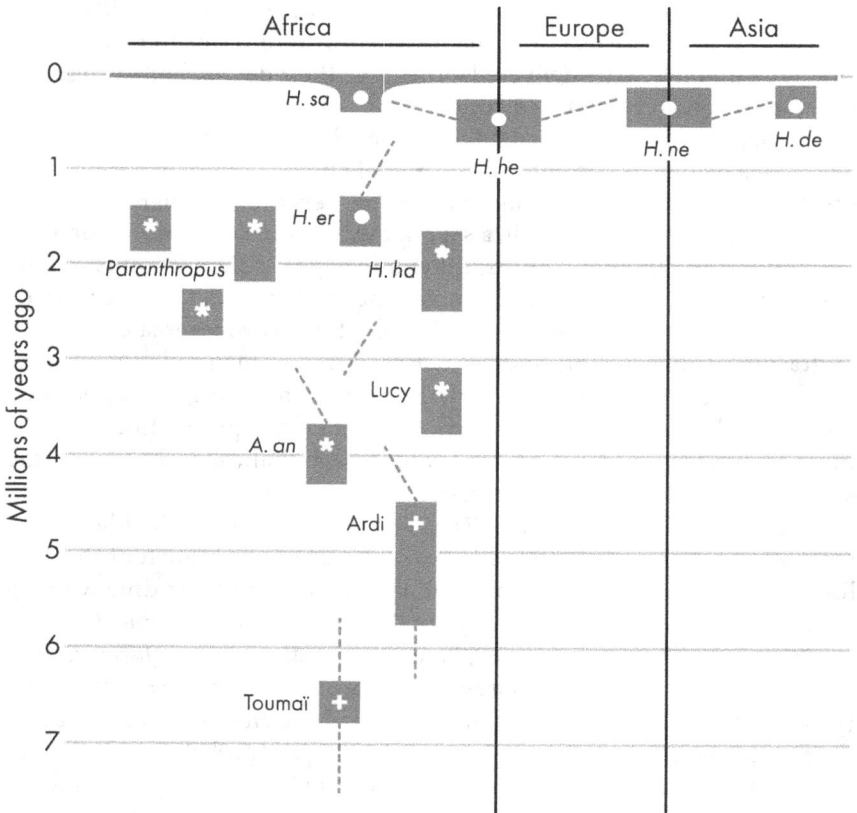

Figure 47. An evolutionary tree of upright walking. Round spots represent species with an essentially modern bipedal locomotion. *H.sa* is *Homo sapiens*. See text for detailed description.

was apparently adapted to living both in trees and on the ground. Toumaï might have shared that same kind of movement and posture and the same kind of environment, represented by "+" in the diagram above: an upright style of living not like that of modern humans but well adapted to the environmental change as forests gradually became patchy and gave way to more open woodlands.

Why didn't Ardi's descendants—and we may be one of them—stay with Ardi's style of locomotion, which seemingly gave the best of both worlds, climbing trees or scampering along the ground as circumstances required? The answer is that Ardi's world changed. According to analyses of ancient lake sediments and of fossil assemblages of land animals and plants, the conditions in Africa three or four million years ago were slowly becoming cooler and drier. Forested areas became patchier, and savannah grasslands began to open up. Here was an evolutionary opportunity, given an appropriate set of mutations that could be subject to natural selection. Some ape species gradually evolved modifications in their behavior, their daily routines. They began trading their green and leafy existence for that of more open woodlands with greater distances from tree to tree, and eventually for the wide-open savannah with its hot sun and open plains, its grasses and sedges. The old forest lifestyle slowly disappeared; treetops gradually became foreign territory. Over the millennia several species of ape, including *Australopithecus* (Lucy and her relatives) and the three species of *Paranthropus* ("*" in the diagram above), replaced the transitory four-footed means of travel by increasingly two-footed ambulation. Our ancestors spent less time climbing through the branches and more time slouching along the ground. The new species extended their range, foraging for grasses and underground bulbs, tubers, and roots to add to their old diet of leaves and fruits. As the millennia passed, the anatomy was slowly modified by natural selection of the appropriate mutations. Teeth changed to suit the dietary change. The hind limbs elongated and the pelvis and hip joints adapted to better support the weight of the upper body. The foot elongated and became flatter, the big toe shifted to a more forward-pointing position, the heel and ankle bones were altered, and the vertebral column changed to accommodate a more upright stance. There was a shift in foramen magnum from the back of the skull to the bottom, in accordance with the increasingly upright posture. The evolutionary changes seen in the fossils of our evolutionary human ancestors in the genus *Homo* gave a more firmly upright posture and more regular walking habits, more similar to ours than were Toumaï's or Ardi's. (We don't feel like calling every kind of upright posture an "evolutionary advance" just because those other styles of locomotion were somewhat similar to our own. In evolutionary terms they were not really advances. Evolution doesn't work that way; individuals never aim for some future outcome. Those who are better adapted, survive more often, and have more offspring simply pass on the genes for their better-adapted anatomy in greater numbers to the following generations.)

Lucy herself, an adult female, was short, about the size of a modern six- or seven-year-old human. Her pelvis, leg bones, knee and ankle joints, were quite like those of modern humans. Her feet were also similar but not identical to ours. She didn't have quite the foot arch that our modern foot has. Her stride was like ours, but her abilities as a long-distance runner were probably limited. For her size, her arms were long and her fingers more curved, indicating a better climbing ability than we have. Ironically, judging by the apparent fractures of her pelvis and her arm and leg bones, Lucy may have died by falling out of a tree.

When savannahs eventually opened up, the populations to which Ardi had belonged gradually evolved away from their previous strict leaf-and-fruit diet to one that included grasses and roots, perhaps also supplemented with mice, squirrels, rabbits, which could be caught at ground level. Eventually large herds of grazing animals appeared. Populations of new apes emerged that were adapted to this new environment. Over thousands of years these apes became more exclusively ground foragers, scavengers and long-range hunters. As evolution proceeded, everything dovetailed to make a new and larger species of ape, one with longer legs providing more efficient travelling on the ground, forward-pointing big toes, and rounded heels, permitting long-distance running. The forelimbs, freed from the tedious task of locomotion, could be used to could carry food back to the home base from long distances and to make stone tools. An omnivorous diet provided better and more consistent nutrition, allowing a bigger and better brain to evolve. This gave the strange new human-like ape the capacity to cooperate more fully with others in the hunt, and to mentally put itself in the mind of the antelope it was tracking.

When *Homo ergaster* (H. er in Figure 47) appeared almost two million years ago, the modern mode of human locomotion (indicated by the round white circle) was in place. (The earlier *Homo habilis*, H. ha, seems to have been intermediate between Lucy and *Homo ergaster*.) *Homo ergaster* was as tall as modern humans and had an anatomy that could support the exclusive use of its two hind legs for locomotion, especially for running and chasing down antelopes and giraffes and buffalos. Evolutionary change modified the length and diameter of the leg bones, the construction of ankle and knee joints and the form of the heel and other foot bones (which absorbed the shock of coming down to earth from mid-stride), as well as the structure and attachments of the Achilles tendon, the big toe and its alignment with the whole foot for pushing off, the structure of the spinal column, the shoulders, thorax, and pelvic bones (to compensate for hip rotation while running), the structure of the skull and the ligament attaching it to the vertebral column, the semicircular canals of the inner ear (for balance while running), and the skin, resulting in loss of body hair and changes in the distribution of capillaries and sweat glands, allowing for evaporative cooling while running in the hot open African plains.

These changes, which appear to have occurred first in *Homo ergaster*, were probably refined as *Homo sapiens* was coming into being. While we may not always behave like *Homo sapiens*, the *Wise Man* that we think we are, we definitely are *Homo cursorius*, the *Running Man* (Figure 48).

However much we now sit at our desks and in our armchairs, long-distance running is what humans were originally built for. Over short distances, a chimpanzee, a horse, a cow, a dog or even a cat can outrun a human being, but over long distances on a straight course a fit human being will run them all to exhaustion.

It is sometimes said that runners have only so many marathons in them. But members of the "100 Marathon Club" have run at least that many marathons, and often more. A colleague of mine has run 20 marathons and also 10 ultra-marathons, 100 miles at one go, and he was still going strong the last I heard.

Figure 48. Three runners, from a Greek terracotta amphora from the fourth century BCE (Wikimedia Commons).

13

The Brain

The Brain—is wider than the Sky—
For—put them side by side—
The one the other will contain
With ease....
—Emily Dickinson, poem 632 (ca. 1860s)

We humans, at least by our own criteria, are the most intelligent and creative of creatures. Signs of it are all around us: airplanes, automobiles, bookstores, cell phones, rock concerts, poetry readings, sewer systems, skyscrapers, supermarkets, roller coasters, manicured flower gardens. We build cities, cultivate the land, domesticate plants and animals for our use, extract coal and oil from the ground and burn it for its energy content, synthesize antibiotics, harness x-rays and magnetic fields for medical diagnosis and research. We write symphonies, novels, poems, philosophical treatises, historical analyses. We put together dance repertoires, build scientific theories, devise political strategies, create systems of laws, and send rockets to the moon.

These days we look for the source of these special abilities in our individually inventive and collectively interacting human brains. The ancients weren't so sure. Aristotle speculated that the brain might be a radiator to cool the head. Empedocles suggested that the source not only of emotions but also of thoughts was the heart, not the head: "Nourished in a sea of churning blood, where what men call thought is especially found—for the blood about the heart is thought for men." Eventually students of human anatomy found the right path, observing of the effects of brain injuries on speech and cognitive abilities.

It is in the area of higher-level brain functions that the human being seems least like a machine and most like a non-physical being that is more than the sum of its gene- and protein-directed parts. In *Song of Myself* Walt Whitman wrote,

People I meet, the effect upon me of my early life or the ward
and city I live in, or the nation,
The latest dates, discoveries, inventions, societies, authors old and new,
My dinner, dress, associates, looks, compliments, dues....
the fitful events;
These come to me days and nights and go from me again,
But they are not the Me myself.

The question of a sense of self beyond the purely physical is a contentious one, about which many books have been written. The mind seems somehow separate and distinct from the physical brain itself. But it is most likely not, any more than the music played on a violin is something separate from the physical instrument.

Brain Development and Structure

In the earliest stages of embryonic development, the nervous system is a relatively simple hollow tube with only minor structural variation along its length. After a couple of months, at a stage when the growth of the brain begins to slow down in other mammals, the division of cells in humans continues apace, especially in the front end, with millions of new cells being added every hour. In the third and fourth months of fetal life the cells of the forebrain continue dividing, and the underlying, slower growing, more primitive parts of the brain become completely overlain by folded layers of nerve cells and supporting cells. This is the cerebral cortex. It continues to grow even after birth, enlarged relative to our body size beyond what is observed in other animals, even in other primates.

After about 18 years of growth, the human brain reaches its adult size: an organ weighing about 1300 grams (about three pounds). The cerebral cortex, containing billions of nerve cells and trillions of interconnections, makes up three-quarters of the weight of the whole brain. The brain is a squishy mass of tightly packed cells; from it arises all human thought.

The anatomy of the human brain is shown in Figure 49. The cerebrum, the major part of the brain, consists of

- the cerebral cortex, which is made up of several thin surface layers ("gray matter") of nerve-cell bodies, the layers being folded into ridges and gullies, the so-called brain convolutions;
- the underlying tracts of nerve-cell extensions ("white matter") that serve as connections between different nerve cells, between different parts of the brain, and between the brain and the rest of the body; and
- several interior nerve-cell aggregations.

Unfolded and laid out flat, the brain's total surface area is greater than that of other apes, adjusted to body size. The two sides of the brain together have a surface area greater than that of two large pizzas. It is also an energy-intensive organ: the brain occupies only about 2 percent of our body weight, but it consumes a full 20 percent of our body's energy intake. With all that feverish activity going on right behind our forehead, it's not surprising that the ancients thought the human brain might be a radiator for getting rid of excess body heat. The large amount of metabolic heat produced by the brain has led to the evolution of a special system of blood capillaries in the skull for keeping the brain cool.

From observations on the effects of brain injuries and from magnetic resonance imaging of conscious people engaged in various mental tasks, we know that different regions of the cerebral cortex, although interconnected, are nevertheless specialized for different functions involved in voluntary behavior. The cerebral cortex is anatomically divided into four lobes on each side of the brain, convenient for describing general locations of different functional regions. In the frontal lobe, muscle movements are initiated, planned, and guided. Also in the frontal lobe, especially in the "prefrontal cortex," the forwardmost part of the frontal lobe, just behind the forehead, are located the mental processes of thinking, analyzing, problem-solving, and decision-making. This part of the brain is enlarged in humans compared to other primates, hence our forward-bulging forehead. Brain cells in the cortex of the frontal lobe, connected with other parts of the brain, contain important parts of a "curiosity network," the cerebral basis of the biological drive behind all of science (and the seat of a large part of the motivation for writing this book). In the parietal lobe, nerve cells receive and process much of the sensory information from the body, including the sensation of pain. The occipital lobe, at the back of the brain, receives and analyzes visual information from the retinas of the eyes. The temporal lobe encompasses the functions of hearing, music appreciation, the understanding of spoken language, and some aspects of memory formation. In the cerebellum ("little cerebrum") at the base of the brain are located brain cells responsible for coordination of body movements, as, for example, in playing the piano or ice skating.

Compared with those of most other mammals, humans have large brains. But our special human abilities cannot be simply a matter of brain size. Elephants and whales have larger brains than humans do, but they don't measure up to humans in ability to solve difficult problems, in technology and language, and maybe not in the complexity of their social relations either. Elephants don't write poetry and publish books on science and philosophy; whales don't compose symphonies and perform them in undersea concert halls; killer whales haven't devised spear guns and fishing

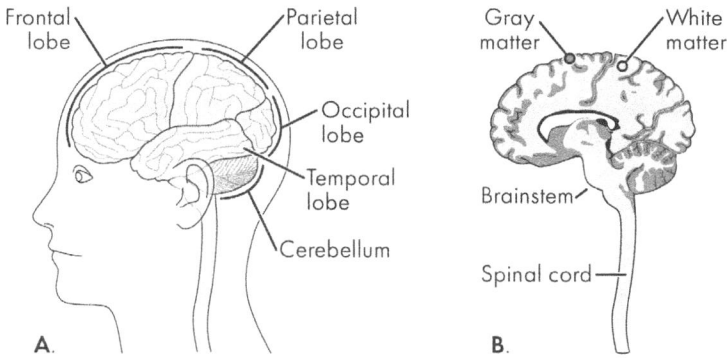

Figure 49. The structure of the human brain. (A) The four main lobes of the cerebral cortex in the left hemisphere (the right hemisphere is its mirror image). The front part of frontal lobe, just above the eyes, is the "prefrontal cortex." (B) A vertical (sagittal) slice just off-center, with the thin cerebral cortex (gray matter) and the underlying nerve tracts (white matter) labeled. The brainstem controls involuntary bodily activities such as heart rate, breathing rate, waking-sleep cycles, blood pressure control, and other involuntary reflexes.

nets as a more efficient means of capturing their prey. Mere brain size doesn't explain anything; we need to identify the particular features of the human brain—the details of its "functional anatomy"—to explain how it works. To account for our special abilities we also need to view the human brain as part of a larger biological system, one that takes account of the evolution of our hands (free from the task of moving the body from one place to another and free to make sophisticated tools), our complex social interactions (allowing for extensive cooperation and competition), our language and culture (providing for the transmission of thoughts and information from one person to another and from one generation to the next).

The human brain is arguably the most complicated object in the Universe. In our studies of it we are like explorers who are proud of having cut through several hundred yards of dense thicket in our search for knowledge about the brain but are still uncertain about whether we're headed in the right direction, and dimly aware that the jungle extends for thousands of miles on every side.

Some Comparisons

The human being is not as large as a whale or an elephant but is still a large animal, as mammals go. Among mammals there is a close correlation between body size and brain size (measured in terms of mass, weight, or volume). You can predict fairly accurately the size of the brain from the size of the body that houses it (Figure 50). A mouse's body is small, with fewer cells, and a small brain, while an elephant's large body has many more cells, and a large brain to match. Larger animals must have larger brains to receive all the sensory information relayed from the body's surface and its inside, and to make sense of it. In order to command the larger-than-average muscle mass, it needs more brain cells to send out signals for movement. To regulate the body's physiology and to coordinate all the input and output, a larger body needs a larger brain. It could hardly be otherwise.

Nevertheless, the primate brain and especially the human brain is larger than it needs to be for these basic functions. Gibbons, chimpanzees, orangutans, and humans all lie above an "average" line for mammals (Figure 50). On this scale our brain stands out as being seven times larger than it needs to be for the size of animal we are.

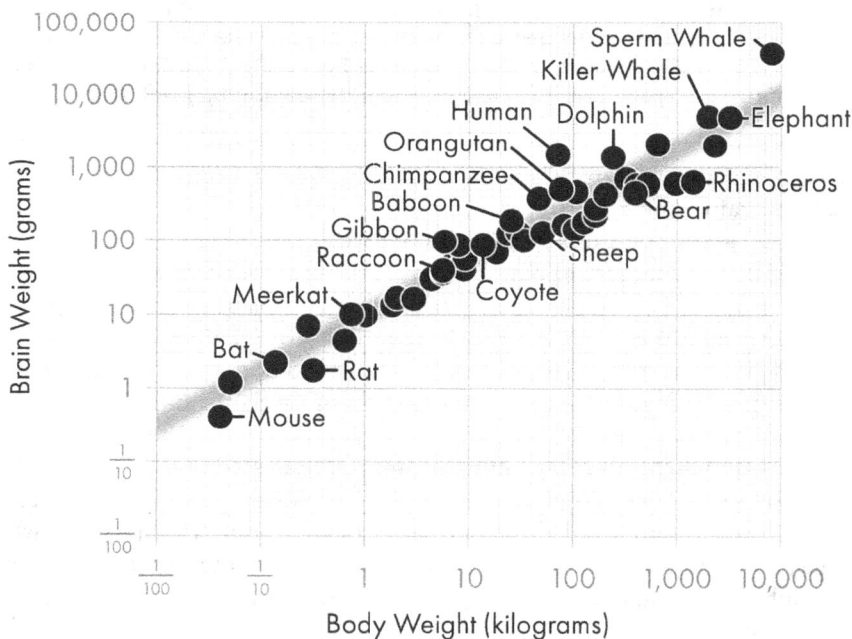

Figure 50. The relation between body size and brain size in mammals. Because the sperm whale is 200,000 times larger than a mouse, both axes have logarithmic scales in order to fit animals of such disparate sizes onto the same graph (graph reconstructed from data in G. Tartarelli and M. Bisconti, 2007, Trajectories and constraints in brain evolution in primates and cetaceans, *Human Evolution* 21:275–287, and other sources).

Brain size by itself doesn't mean much; it's too crude a measure of what a brain might be capable of. The brains of two different species could be the same size but have different capabilities because they are wired differently or have specialized regions devoted to different functions. The brain of the bottlenose dolphin has enlarged regions devoted to auditory functions and echolocation. Portions of the elephant brain specialized for cognitive abilities are smaller than the corresponding regions in the human brain, but regions specialized for processing and coordinating sensory information and movements are relatively larger than those in humans, perhaps because of the huge skin area and the sensitivity and versatility of the elephant's trunk. All mammalian brains have regions specialized for receiving and processing visual input as well as input from the senses of hearing, touch, smell, and taste, for coordinating body movements, for memory, for problem-solving, for organizing remembered experiences, for analyzing events in the environment and adjusting behavior accordingly, for self-control, for planning and decision-making, for learning, for language, for emotions.

The assumption that brain size is related to intelligence needs to be supported by evidence. If the assumption were true in all cases, then since men's brains on average are 10 percent larger than the brains of women, men should be 10 percent more intelligent than women. This is a proposition for which there is no good evidence. Besides, men's bodies are on average about 20 percent larger than women's bodies, so men's brains are not even keeping up with their body size.

The size of the brain and the total number of brain cells it contains might not, in principle, be directly correlated with intelligence. Bigger brains might simply be necessary for running bigger bodies. There might be only some small, specialized region of the brain where cognitive ability is located, a region that does not correlate with overall brain size. There might be no correlation between brain size and cognitive ability if the latter is primarily a matter of how the brain is wired, not how big it is or how many cells it has.

In spite of these theoretical reservations, the evidence is that in fact, on the major avian and primate branches of the grand evolutionary tree, the bigger the brain, the greater the cognitive abilities.

Parrots and crows are notable for their communication skills, their ability to solve problems and use tools, their understanding of cause and effect, their ability to anticipate the future behavior of others, and to plan for the future. And it turns out that parrots and crows have more brain cells than other birds.

From small marmosets and lemurs to medium-sized monkeys to the larger baboons and gibbons and chimps and gorillas, non-human primate can be ranked according to their ability to solve problems set by laboratory researchers. The problem-solving tests are usually related to finding a food item or getting a food reward; they measure something that can reasonably be reckoned as basic intelligence. They include tests of the animal's ability to sort objects by size or color in order to receive the reward; to use a subtle procedural cue to find a piece of candy that had been secretly transferred from one box to another; to choose an appropriate stick, and even modify it, for pushing a food item out of a cylinder; or to choose an object differing in some way from other objects in order to find the reward, requiring a concept of "same-different." Different species can be given the same tests, yielding scores that allow a comparison of different species' basic intelligence. What correlates most closely with the intelligence rankings is the absolute brain size of the animal. Absolute brain size makes a difference; it isn't just some subtle wiring difference, or a special "intelligence region" in the brain. Across species, purely and simply, animals with bigger brains are smarter. A bigger brain means more brain-cell interconnections, better information processing, and a better ability to analyze what's going on in the world and how to plan accordingly.

Aspects of Brain Evolution

The evolutionary trend over time among the primates was toward larger body size. That in itself means larger brains. But as illustrated by the gibbons, baboons, chimpanzees, and orangutans in Figure 50, the evolution of primates as a group already entailed having larger brains for their body size than other mammals. This may have been the result of evolutionary adaptations accompanying the transition from a nocturnal to a diurnal lifestyle with its greater reliance on visual exploration and manipulation of the environment. Our primate ancestors already had a head start in brain enlargement.

Another factor in the evolution of the brain was a change in diet. The other living great apes—orangutans, gorillas, bonobos, and chimpanzees—subsist primarily on fruits, seeds, nuts, and leaves. The diet of many of the ancient apes that preceded us, to judge from the structure and wear of their fossilized teeth and the chemical composition of their teeth and bones, was similar. Now, the human brain is one of the most energy-intensive organs in the body; in metabolic activity it rivals the liver, kidneys, and heart. It occupies only 2 percent of the mass of the human body, yet with all its ion transport, electrical activity, and continual release and uptake of neurotransmitters, it utilizes a full 20 percent of the body's total energy budget. Beginning probably with *Homo habilis* two million years ago, and certainly by the time *Homo erectus* had appeared one and a half million years ago, members of the genus *Homo* started adding meat to their diet by means of scavenging and hunting. For brain development, this more efficient source of energy and protein, especially with the use of fire for cooking, made all the difference. It allowed a greater evolutionary brain expansion in the human line. Meanwhile, our great-ape vegetarian cousins with their equally large or even larger bodies did not evolve a brain size that quite matched their body size.

As shown in Figure 51, from early in the hominin line (the species in the genus *Homo* and its evolutionary ancestors) there was an increase in both body size and brain size. Both *Australopithecus afarensis* ("Lucy"), from over three million years ago, and *Homo habilis*, from about two million years ago, stood about four feet tall, while more recent fossils such as *Homo erectus*, as well as modern *Homo sapiens*, approach six feet in height.

Adding to the hominin's evolutionary increase in brain size, besides the increasing numbers of nerve cells, were the nerve cells' axons and dendrites, the cell extensions connecting nerve cells to one another. The large size of the modern human brain is therefore the consequence of both the evolutionary increase in the number of nerve cells, as expected for a great

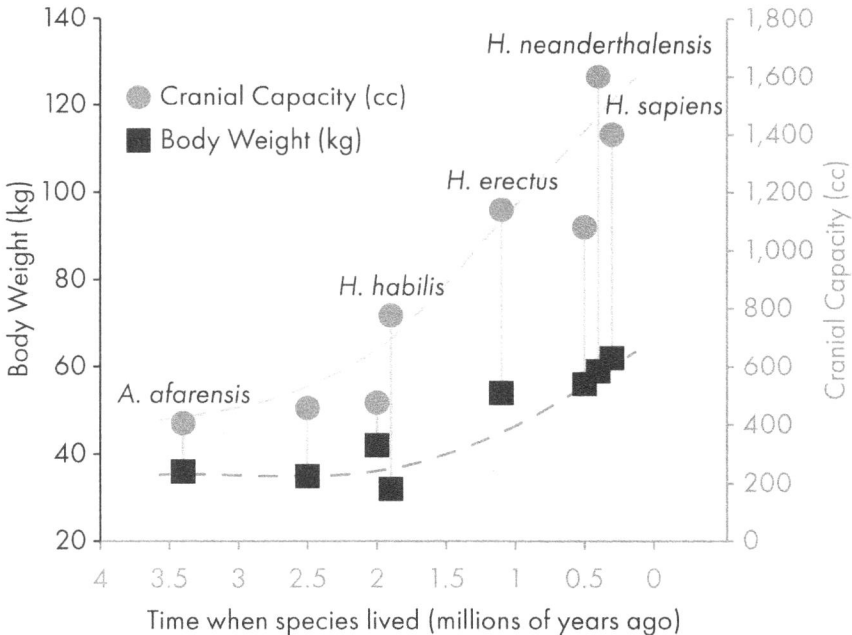

Figure 51. Evolutionary increase in body size (squares) and brain size (circles) in early humans and some of their ancestors. In the genus *Homo*, relative brain size has increased more than body size (graph constructed from information in the National Museum of Natural History, Smithsonian Institution's Human Origins Program).

ape our size, and the greater size of the individual nerve cells, taking into account all the axons and dendrites (Figure 52). Both these features of the brain result in greater numbers of brain-cell interconnections, which may help to account for its great cognitive capacity.

The cause, or causes, of the evolutionary increase of brain size and cognitive capacity is uncertain. Maybe climate fluctuations or the patchiness of the environments in eastern Africa that early apes had to deal with drove some populations extinct; the survivors were the ones who happened to experience gene mutations giving them an ability to survive under such conditions. We might be their descendants. Perhaps there was something about superior cognitive ability that was more directly related to reproduction; for example, both males and females might have been able to detect superior problem-solving abilities in their prospective mates, choosing them over others to have offspring with. If such abilities were dependent on brain-enhancement genes, those genes could have been passed on to more and better-cared-for offspring than other parents could produce. As one generation succeeded another, those genes would come to characterize a whole new population of apes—our ancestors.

Even after our species first appeared on the Earth, there were further changes in store. Evolutionary processes do not stop when a new species comes into being. If the whole evolutionary history of the genus *Homo*—the

EVOLUTION OF BRAIN SIZE FROM EARLY PRIMATES TO HUMANS

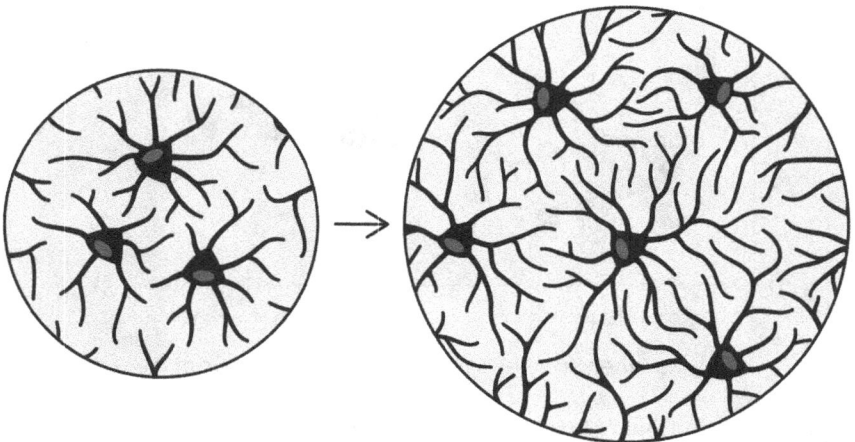

Figure 52. A schematized diagram showing the cellular basis of the enlargement of the primate brain over evolutionary time, involving both an increase in brain-cell number and an increase in brain-cell size.

approximately three million years from *Homo habilis* to modern-day *Homo sapiens*—is represented as the length of this sentence, then our species first made its appearance at the fifth word before the end of the sentence. From that point on, it appears from the archeological evidence that *Homo sapiens* mostly muddled on, doing things the way they had been doing them for over 200,000 years, living selfishly and precariously in small groups, making the crudest of stone tools, scavenging and hunting when they could. Then sometime in the last word of that sentence—40,000 or 45,000 years ago—new cultural advances began to emerge. Stone tools began to be more carefully made and more sophisticated, animal bones were added to tool-making kits, along with technical innovation in general. Clothing and bead necklaces were fashioned, cave art and small carved figurines appear in the archeological record, more systematic and cooperative hunting was practiced. The ability to learn from the behavior and the teaching of others, rather than learning only by individual trial-and-error, probably escalated during this period. Language and symbolic modes of thought (personal adornments stand for social status, vocal sounds stand for animals), as well as social systems of morality, justice, and religion may have had their origins around this time.

If this interpretation of archeological findings is correct and not prejudiced by a patchy and incomplete archeological record and by archeologists' having concentrated on findings only in particular locations—this bias remains a possibility—how is this supposed "great leap forward" in humanity's cognitive ability to be explained? There are two possibilities in contention. Both have their "sudden" and "gradual" versions. One is that, without any increase in brain size, there were genetically based changes in how the human brain operates—a rewiring that opened up new channels of communication between brain cells. The other is that the change was not genetic but social and cultural. Without having undergone any fundamental brain change, humans of 40,000 years ago may have begun to interact socially in new ways, learning from one another, trading their technologies, adapting to an altered social landscape. The distinction between these two scenarios is useful conceptually, but the cognitive advance could in fact have been a combination of both. Gene-based brain changes could have generated more social interactions, and these in turn could have provided the background in which certain kinds of genes were favored over others.

Brain Function

We have made a beginning in understanding our large brain by describing what it consists of. We still need to find out the details of how the brain's different parts work and their functions translate into the human brain's special abilities.

We share with other vertebrates the brain control of basic functions such as defensive responses to danger and the regulation of breathing, blood circulation, wake-sleep cycles, swallowing, urination and defecation.

Any animal's behavior depends on the activity going on in its brain in the context of its particular anatomy, lifestyle, and ecology (you're not likely to find a polar bear tending a vegetable garden). Although in us humans *thought* seems often to precede action, we have general patterns of brain function and behavior that are evolutionarily very old, shared by essentially all vertebrates. The brain of the human being, like all vertebrates, takes in sensory information, processes it, stores it, and reconstructs from it a serviceable semblance of reality. The mental world views of a bat, which depends on echolocation, and that of an electroreceptive fish, which depends on sensing electric fields in its environment, are different from ours, which depends largely on vision. But those other animals do have a worldview nonetheless, constructed by their brains.

Other human brain functions are newer, the hallmarks of behavior which we share with many other mammals: a highly developed memory, the capacity to learn from experience, senses of pleasure and satisfaction. Another mammalian trait is the playfulness of the young, which is part of learning how to deal with the world; it can be observed in young gorillas, wolf cubs, hyenas, skunks, elephants and rhinoceroses and, of course, humans.

Then there are other brain functions that are highly developed in us humans but absent or developed to a lesser degree in other primates:

- The social functions: the ability to put oneself mentally in the place of another, to divine another's intentions, knowledge, and beliefs; to help others; to cooperate to an uncommon degree in common endeavors; to become a follower of a leader;
- the ability to unravel complex networks of causes and effects;
- the ability to plan ahead, to imagine more than one solution to a problem, to imagine events that have not yet occurred;
- feelings of hope, despair, and boredom;
- a sense of fairness and justice;
- the ability to understand and use complex language and number systems;
- the ability to create representations of the world and our feelings in painting, music, and literature;
- attaching high importance to mental activity to the exclusion of all

else, sometimes even to the extent of overpowering any concern for physical well-being; and

- self-awareness, consciousness of what's going on in one's own brain and the ability to report on it.

Exactly how all this emerges from our brain—a collection of cells that does not look all that different from that of other animals—is a mystery. Even vision, a function we share with many other animals, is mysterious in its workings. I see the computer screen in front me, I register that it is rectangular, that its background displays an impressionistic beach scene with blues and greens and splashes of red, brown, and gray. I "see" it. But what does that mean? The first stages of the process are more or less understood. Light of different wavelengths hits the retina and is absorbed by photoreceptive protein molecules in specialized retinal cells, is transformed into electrical signals that are first integrated by other nerve cells in the retina, and then are passed on to the visual cortex in the occipital lobe at the back of the brain (Figure 49) and are further processed there before being transmitted to other parts of the brain. What the visual cortex does with the electrical signals from the retina, and how it and other parts of the brain together produce the subjective experience of "seeing" is the mystery. It's not just a simple pathway from retina to brain, like a reflex, for not only do I "see" what's in front of me, but I also *know* I'm seeing it. Francis Crick and his colleagues have made a start in trying to understand all this, as he described in his book *The Astonishing Hypothesis*—the proposition that everything we learn from birth onwards about making sense of the world around us, everything we experience subjectively, comes only from little watery assemblies of proteins and lipid membranes interacting with one another.

Another unique feature of the human mind is the importance attached to mental life over physical life. People often devote their lives entirely to art, literature, music, science, religious beliefs, political ideals, or particular idiosyncratic goals. Why risk one's life climbing Mt. Everest, or cave diving, or investigating venomous snakes in the jungles of Southeast Asia? People are even willing to die for an idea or a belief. Sir Thomas More, by refusing to acknowledge Henry VIII as supreme head of the Church of England, was convicted of treason and beheaded in 1535. In 1600 Giordano Bruno was burned at the stake for expressing his heretical religious views. Dietrich Bonhoeffer was hanged in 1945 for opposing the Nazi regime. There is nothing like such behavior elsewhere in the animal world. The explanation of this aspect of the human psyche, depending on the particular case, may lie in some blend of curiosity (in cases of exploration), social allegiance, and real or imagined social reward.

14

The Sexy Beast,
Part II

*She must be torn asunder for life to come forth, yet still
they were one flesh, and still, from further back, the life
came out of him to her ... their flesh was one rock from
which the life gushed, out of her who was smitten and rent,
from him who quivered and yielded.*
—D.H. Lawrence, *The Rainbow* (1915)

A secret peephole for videotaping a naked woman in a hotel room, images disseminated on public media, a police investigation, an arrest, a jail sentence, a lawsuit against the hotel, a countersuit, months of legal wrangling, an eventual award of millions of dollars—could there be a better example of the human obsession with the naked human body and sexuality? Anti-obscenity laws, public-decency statutes, the hijab, female genital mutilation, fig leaves in paintings and on statuary, widespread social and religious prohibitions against displays of human nakedness and human sexuality in general—all bespeak the nearly universal human obsession with sex.

Poetry, songs, novels, art, music, dance, and dreams are pervaded by references to sex. Gender is assigned to inanimate objects without rhyme or reason ... *bridge* is masculine in Italian (*il ponte*), feminine in German, (*die Brücke*); *chair* is feminine in Italian (*la sedia*), masculine in German (*der Stuhl*). Sexual images and sexual innuendos are used to sell motorcycles, chocolate bars, and soap (Figure 53). Human sexuality is the backdrop and context of TV dramas and quiz shows. Sexual attraction, sexual jealousy, and unrequited love are the basis for the exploits of Greek gods and goddesses and for the stories in plays, operas, and fairytales. Gender differences are accentuated in ballroom dancing, the *pas de deux* of ballets, pair skating, and the fashion industry. Much of the world's literature has male-female relations as theme or background. Special rites of passage surround the sexual coming-of-age of both females and males. Forming a long-term bond between a man and a woman (or sometimes among more than two people) is often marked by a special, often religious, ceremony.

Our mating behavior seems, at first sight, to be governed by social conventions and norms that we could overturn at any time. But there are prevailing, species-wide practices that without thinking we accept as normal, when by ordinary standards of primate biology, they are exceptional.

To the evolutionary biologist, the "human mating system" means not only the attraction between the sexes, the dating game and how mating partners are chosen, copulation practices, and marriage rituals, but also includes the timing of sexual maturation, ovulation cycles, anatomical adaptations for childbirth, the spacing and maturation of offspring, the roles of the parents in caring for the young, and the duration of reproductive life. Some of these are covered in the following paragraphs.

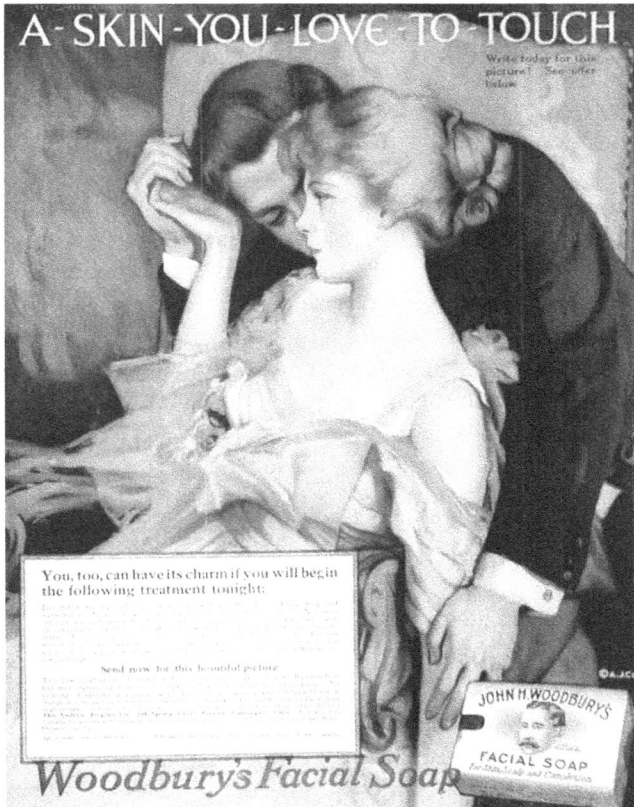

Figure 53. "A skin you love to touch." Using sexual allure to sell soap (1916 American ad for Woodbury soap; painting by Mary Greene Blumenschein).

We humans have a unique mating system. In many ways it is unlike that of most other primates. Our copulation is private. We enjoy sex for pleasure at almost any time, not just when there's an egg ready to be fertilized. Rather than spacing out our offspring so that the latest one comes along only after the previous one has become independent, we have several immature children running around at the same time. They all require many years of intense care and attention in order for them to grow up. Unlike the case with the other great apes and most other primates, the human mother often has direct help with the birth process itself, and almost always with the rearing of offspring. This is often provided by the biological father, but sometimes by other family members or by the wider community. Human females are unusual in several ways. They show few overt signs of their fertile period. They have orgasms. When mature, they have prominent breasts even when they're not lactating. They live well beyond their reproductive years.

Our species has other traits unusual for primates: we avoid incest not by having the members of one sex leave their birthplace, as many other primates do, but by a psychological means, the incest taboo. Males tend to compete for females by demonstrating their mental prowess and ability to garner resources, and less by direct physical combat as deer, zebras, giraffes, and gorillas do. Males of our species don't have the physical traits that characterize the males of the other great apes. Unlike male chimps, gorillas, and orangutans, human males are not much larger than females, and their canine teeth are not viciously long. In all these ways, humans are unusual.

These features are not independent of one another. The threads of anatomy and instinct, male and female, are evolutionarily interwoven to create the warp and woof of the tapestry of human life.

In our sexual anatomy, physiology, and behavior the general principle, as usual, is that our oldest features are the ones we share with many other animals. Newer ones we share only with some. The newest belong uniquely to our own species.

Some of our anatomical features and behavioral traits related to reproduction—penis and vagina, the pleasure of copulation, external testes, prominent mammary glands, hidden fertile periods, menopause—have been inherited from our distant 100-million-year-old mammalian ancestors; all modern mammals share them. Some must be more recent, being found only in primates, and dating from about 70 million years ago. And some are uniquely human, not older than our earliest human ancestors, who first appeared about two or three million years ago. The final result is the patchwork pattern of anatomy, physiology, and behavior that would be expected from the branching course of evolution. The trunk's branches have general features in common; one branch's twigs share traits peculiar to that branch; and each twig has characteristics unique to itself.

Described below, in approximate order of evolutionary age from older to younger, are a few of the anatomical, physiological, and behavioral features that have arisen over the course of human evolution.

The Penis and the Vagina

The main thing is to get the sperm to the egg. Different animal species have their own ways of doing it, depending on their lifestyle. From the human point of view, the bedbug and the pseudoscorpion (a tiny relative of spiders and scorpions) win prizes for pure bizarreness. The male bedbug practices "traumatic insemination." His sperm-transfer structure is like a miniature hypodermic needle. He pierces the female at almost any place on her body and injects the sperm directly into her body cavity, leaving a wound. The sperm then find their own way to the ovary and the eggs.

In the pseudoscorpion *Serianus*, the male has no direct sexual contact with the female. When he senses that females are in the vicinity, he searches for a suitable space underneath a pebble or piece of wood and secretes a small stalk, at the top of which he attaches a small packet of sperm (technically called a *spermatophore*). Then, using a gland at his rear end, he spins silken threads in two nearly parallel vertical zigzag rows that are farther apart at one end and closer together near the spermatophore. The female is led down the narrowing path directly to where the spermatophore is waiting. She positions herself over the spermatophore, takes it into her genital opening, and moves away, leaving a bare little stalk behind. (How different human society might be if humans adopted this method of courting and fertilization. Instead of a face-to-face date that ended up with the two of them in bed, the man might simply leave instructions at the woman's door as to where she might find his packet of sperm, and then be on his way.)

Most aquatic and semi-aquatic animals—fish, amphibians, and the majority of marine invertebrates such as jellyfish, sea urchins, and clams—have no need for a penis or a vagina: females and males just release clouds of eggs and sperm into the water. Once evolution had produced animals that spent all their lives on land, a method had to evolve to have eggs and sperm meet in a wet environment: internal fertilization. Theoretically, eggs could have been transferred to the male for fertilization inside *his* body, and he would then lay the eggs or carry the growing embryos. But sperm are the motile gametes, so it's better done the other way around: sperm find their way to the eggs inside the female. She's the one who determines what's to be done with them next. Aside from her being the egg-producer, her role defines her as to what it means to be female. A tube for conducting the sperm cells out of the male and a female channel for receiving them were natural developments. Aside from the egg- and sperm-making organs themselves, the penis and the vagina may be among the oldest anatomical structures to grace the mammal. (The words come from the Latin *pēnis*, meaning "tail," perhaps referring in some ancient language to either one of those bodily appendages; and the Latin *vāgīna*, a sheath or scabbard.) These structures likely first arose more than 200 million years ago in the common ancestors of reptiles and mammals.

The males of most modern reptiles have penises. A male turtle has one, and so does a male crocodile. Male snakes and lizards, following their embryological rules of bilateral symmetry, have two, one on each side. Depending on the species, they use first one and then the other in successive copulations, transferring sperm to females first from one testis and then from the other.

In the majority of modern birds, who are evolutionary descendants of reptiles, males don't have penises. Both the male and female have an opening for waste, the cloaca (Latin for *drain* or *sewer*), and this chamber also serves the male for the transfer of sperm and the female as a passage for receiving them and for the eventual laying of her eggs. Sperm transfer occurs when the mating pair press their cloacas together in a cloacal kiss. Ostriches, emus, ducks, geese, and swans are exceptions: they have penises. Swan anatomy was undoubtedly known to the ancients; it was no accident that the Greeks had Zeus decide to come down in the form of a swan to copulate with Leda.

The mammalian mode of reproduction involves more intimate contact. The mammalian intromittent organ, the penis, is an ancient feature found in all mammals. Its anatomical details vary, each species having evolved its own peculiarity in matching male and female structures, according to the species' mating habits. Sensory nerve endings in the penis and the vagina are hooked up to pleasure centers in the brain. Many male mammals have a baculum, a bone that stiffens the penis. It is present in bears, seals, raccoons, skunks, weasels, wolves, dogs, cats, rats, and bats. It is also present in most primates, including our close great-ape relatives the chimpanzees, gorillas, and orangutans. But it's missing in humans; it seems that our species has lost the baculum over the course of our evolution; human males rely entirely on a short-term hydraulic blood pressure system to maintain erections.

The baculum, in those species where it occurs, may help the male maintain a copulation session for a long time. Copulations in wolves and weasels may last for a couple of hours or more. Human copulations, not counting foreplay, typically last just five or 10 minutes. It may be that in humans there evolved a mating pattern of short copulations (not requiring a baculum) but more frequent ones. This would have promoted a high likelihood of fertilization, at the same time reinforcing a male-female bond because of the frequently repeated bouts of pleasure—something that might have been helpful during the long years of raising helpless progeny. The absence of a baculum in humans might therefore be an example of a recent evolutionary modification—a loss, in fact—of an old structure that most other primates possess.

Human males might be proud to know that their erect penises, averaging over five inches in length and about one and a half inches in diameter, are larger than those of chimpanzees, gorillas, or orangutans. But, evolutionarily speaking, it could be mostly women who are calling the shots. Human females might have evolved a birth canal to accommodate the human baby's brain. Over the course of evolution human males found their penises adjusting to the wider vaginas. The size of the human penis, as a thoughtful person might put it, could therefore be an indirect consequence of the relatively large size of the human brain. Over the long course of evolution, all the parts fit together.

External Testes

Most men accept things as they are, grit their teeth, and bear up. But is there a male (or at least a male biologist) who, after being kicked in the groin on the rugby field, or falling straddled on the hip bone of an opponent on the basketball court, or slipping and falling astride the balance beam, hasn't asked himself, "Why are the testes placed so as to be so susceptible to such mishaps rather than being sensibly tucked away inside, the way girls' ovaries are?"

In both male and female embryos, the primordial gonads are first formed inside the abdominal cavity near the kidneys. After a couple of months of development, testes and ovaries become distinguishable. Starting at around three months and finishing just before birth, an out-pocketing of the lower front abdominal wall develops—the scrotum—and the testes are dragged down into it by the shortening of ligaments. The complexity of the process suggests it's *for* something; that is, that in the mature male there is some evolutionary advantage to having the testes located in an external scrotal sac.

In many mammals, normal sperm development requires cool storage at a temperature five to ten degrees lower than the body temperature. The location of a testis in a scrotal sac outside the body provides those conditions. But it's not clear why evolution didn't produce sperm cells that could develop in a warmer and safer place inside, at normal body temperature. Could it be that sperm cells need to be shocked into activity by being suddenly moved from a cool testis into a warmer vagina and uterus?

Or it might have been a matter of pressure. Animals that move slowly or smoothly—whales, dolphins, seals, sloths, elephants and moles—have internal testes; the testes stay where they first developed in the embryo and fetus. But animals that run, jump or gallop—many primates, including humans, as well as dogs, cats, bears, horses, antelopes, rabbits, rats, mice, and kangaroos—have external testes. In spite of the disadvantages, testes located in a scrotum might not be subject to sudden and potentially damaging intra-abdominal jolts of pressure.

Not all observations fit this generalization. Shouldn't those great hunters, the female lions, cheetahs, and cougars have external *ovaries* for the same reason? And in spite of their slow-moving lifestyles, koalas and giant pandas have external testes. Other factors must be at work. One of those factors might be the slow creeping pace of evolution. Koalas belong to a group of marsupials that also includes the kangaroos, the koalas having branched off from the kangaroo family millions of years ago. Similarly with the giant panda, which is a kind of bear. It belongs to the great order of carnivores, most of which are active hunters. The giant panda, however, is slow moving and does not have to chase down the bamboo shoots and leaves it lives on. Therefore, some of the anatomical traits of the koala and the giant panda, including their external testes, may be a holdover from their evolutionary heritage, not an adaptation for their present way of living. Nor would humans be immune from the slow pace of evolution. Since the rise of agriculture about 10,000 years ago, most human males, except perhaps in the African Kalahari, no longer run long distances chasing down their antelope prey, yet they still have external testes. Ten thousand years is a short time for evolution. It's easy to imagine, so slow is the evolutionary process, that human males will look pretty much the same thousands of years from now, if our species survives that long.

How Human Brain Size Influences the Mating System

Two or three million years ago the climate in Africa began to change. The wide grassy savannas opened up and became populated with large numbers of grazing animals—antelopes, zebras, wildebeests, gazelles, impalas, and others: meat on the hoof. Some apes who were tentatively beginning to abandon the forest and woodland lifestyles gradually evolved an upright stance, with hind limbs capable of long running strides, forelimbs that could be used for tool manipulation and spear-making, and more sweat glands in an increasingly hairless skin. All these adaptations gave these apes some new advantages in survival and in being able to produce more offspring than their earlier forest-dwelling ancestral apes could manage. They could run down animal prey over long distances, add more energy-laden meat to their diets and thereby feed the growth of larger brains, which could be used for figuring things out, for planning, and for communication. These apes and their descendants flourished and multiplied. Along with these rich evolutionary gifts, however, came some disadvantages. For one thing, the pelvis narrowed while the cranium enlarged. There consequently arose the obstetrical dilemma: how to manage the birth process? The evolutionary channel that opened up was to give birth early, before the brain was fully developed and when the baby's head was still small enough: let the brain finish most of its development *after* birth. Problem solved. But a new dilemma arises: the infant is so *helpless*, and for so many years! It needs so much feeding—so much care, so much protection, so much breast milk. The problem was compounded by another "good" reproductive trait, a short birth interval. With frequent sex and without birth control, the mother popped out a new baby every couple of years, or even more frequently. This is the shortest interval between births of any of the great apes. Two or three sucking mouths all had to be fed at the same time. And for six or eight or more years many more or less helpless bodies had to be given continuous care and protection.

The human being is not like the chimpanzee, who travels about on all fours much of the time, has a relatively wider birth canal and a relatively smaller brain, an easy birth, more rapid maturation of the young, and longer intervals between births. The chimp mother therefore has a relatively easy time of it and can raise her young by herself. On the evolutionary scene there arose in humans some physiological and behavioral modifications that ameliorated the situation, allowing human reproduction to proceed apace. All were aimed at keeping the male around, creating a longer-lasting bond between the mated pair: the result was that there were now two adults, rather than just one, who provided the years of necessary childcare.

Conspicuous Mammary Glands

The evolutionary precursors for the milk glands of mammals likely were skin glands that pre-mammalian animals used to produce secretions to protect their recently laid eggs from drying out. Glands that produced true milk probably evolved somewhat later when gene modifications made those secretions more nutritious. Other gene changes localized the ventral skin glands to particular places and made them more productive and more responsive to female hormones. The young of our pre-mammalian ancestors probably began to be nourished in this way at least 200 million years ago, long before mammals themselves had appeared on the Earth.

Humans are unusual among mammals in the prominence of mature females' mammary glands, well beyond what's necessary for the nourishment of the young (Figure 54). The females of other mammals have plenty of glandular milk-producing tissue without such eye-catching swellings. Female orangutans, gorillas, and chimpanzees have a degree of breast enlargement when they are nursing their offspring, but nothing quite like the permanent part of mature human female anatomy. Besides, most of the natural breast enlargement of human females is the result of fat deposition, not extra milk-producing glands. Fat deposition in the thighs and buttocks of a female might indicate a good supply of nutritional reserves, but why in the breasts?

In human females the enlargement of the breasts is not primarily a sign of pregnancy as it is in other primates but an advertisement of sexual maturity. It is a sexual symbol. It seems to be related to humans' readiness for sex at almost any time, rather than just during the females' fertile periods. To varying degrees this is true in all human cultures. In the film *Men in Black II* the enemy alien Serleena notices how easily distracted men in power are: "Silly little planet," she says. "Anyone could take over the place with the right set of mammary glands." That female breasts are a sexual symbol accounts for the widespread practice of covering them in public. Their sexual symbolism is a specifically human trait. Imagine a bull, a male elephant, a male dog, or a male chimpanzee nuzzling the mammary glands of the female of their species during their mating. That kind of behavior we would view as bizarre. But for us humans it's a normal part of the mating ritual. How did this situation come about?

It may be that over

Figure 54. The prominence of the mammary glands of the human female is not just for milk production but is a sign of sexual maturity (drawing by Carl Newman, 1858–1932, Smithsonian American Art Museum Collection no. 1967.63.163).

evolutionary time our male ancestors gradually came to mate with females who happened to have had more than the average amount of fat anywhere in their bodies, including around the mammary glands. It was a potential sign of the females' good genes and general good health. Genes providing fat deposition in particularly prominent places, especially in animals that walked upright, might have been passed on to greater numbers of offspring. The smaller-breasted females presumably had fewer matings. Over evolutionary time female anatomy would have responded to male preference. And the male preference itself might have arisen because the males, without conscious thought, took visible fat deposition as a sign of good maternal health and good potential for raising his offspring. With time, genes constructing male brains with such a preference ended up being passed on to more offspring. Ancient human populations would have gradually shifted over to having more and more females with prominent breasts and more and more males with a preference for them.

This evolutionary account can also be described in terms of "goals" and "choices," even though we know that evolution has no goals. If new genetic mutations happen to result in better survival or more offspring, then the new genes probably will eventually replace the old. The creatures in question—humans—don't make conscious evolutionary choices, but the story can be described as if they do. The male's "goal" is to get as many of his genes into the next generation as possible. He can best do this by choosing especially healthy and fertile females for mating purposes, the ones with the best chances of raising his offspring to reproductive age. Wide hips and ample breasts could be indirect but nevertheless reliable signs of good female reproductive potential. A female with those characteristics is the kind of female he should choose. On the female's side, it pays her to advertise those attributes—along with a narrow waist and a beautiful symmetrical face—in order to attract as many males as possible. From among her array of choices she can select the strongest, the most caring, the most resourceful, the most successful male. This will maximize her chances of successfully getting *her* genes into the next generation. A man's interest provides the evolutionary drive for the woman's anatomical changes, and her evolutionary advertising of them drives the man's attraction.

Many questions attend this rather simplistic story. If it's true, shouldn't it apply more generally? Why didn't other primates—gibbons, monkeys, baboons, lemurs—follow the same evolutionary pathway? Shouldn't the male's assessment of reliable signs of female health, such as fat deposition around the mammary glands, have been operating in the evolution of these other primates too? Maybe not, if other factors were also involved in the specific evolutionary development of the human species' sexual interactions: upright stance, so that the mammary glands in our female ancestors would have been more readily apparent; lack of other obvious signs of the female's sexual maturity and readiness; and increasing reliance of the male on visual signs from the female. These factors may have been less salient in other primates.

There are, of course, physiological, social, cultural factors that might have influenced the whole system. But the evolutionary-genetic side of the story is dominant. It might seem as if the strings of men and women, without their having any conscious awareness of what's going on, are being pulled by an invisible, capricious, mischievous yet perhaps ultimately beneficent god. But whatever strange and wonderful traits may arise in hookworms, hawks, or humans, it is only survival and reproduction that count in the long-term continuation of the species.

Sex and Pleasure

To want something, to like something—these are propensities of the mammalian and especially the human brain. Wanting and liking have invaded all aspects of human experience. Why else would I bother to write a love poem extolling the beauty of my beloved, when the pleasure of contemplating beauty need not come into the mating ritual at all? Why else would I seek to profit by the instruction in sexual techniques provided by the *Kama Sutra*? Is not the natural act driven by instinct alone all that is needed for human reproduction? The basic instincts have been covered over with a special wiring system for the more conscious processes of *wanting* and *liking* (and their opposites, *disdaining* and *disliking*) as guides for action. Instead of being automatically triggered by instinct alone, our behavior is filtered through the brain's pleasure-displeasure system. Genes and hormones guide the development of the wiring of the male and female brains for sex as well as for the sex organs. Genes and hormones write the script and set the stage for mature sexuality; the backdrop is sexual desire. The sex act itself is the action onstage, accompanied by its intense but short-lived pleasure—*la petite mort*, the little death, as the French call it. Sex is pleasurable without our ever consciously willing it to be so. It is the pleasure—a brain function—that smooths and paves the road from mindless copulation to the joining of egg and sperm.

The egg and the sperm are little cellular machines performing their allotted functions. In fusing together, they are not experiencing pleasure. Nevertheless, we could imagine that Mother Nature, aiming to perpetuate the species, in her wisdom has provided the whole animal with sexual desire, sexual pleasure, and sexual satisfaction. The syllogism runs like this:

having sex leads to orgasm; orgasm is pleasurable; therefore, having sex is pleasurable; if it weren't, the species would die out.

Humans get enjoyment out of the sex act whether or not sperm and egg ever meet. As with birthday presents, it's the act of giving and receiving that gives pleasure, not the gift itself. Since the sex act in many of the "lower" animals seems to take place purely mechanically, at what point in evolution did the pleasure centers in the brain become involved?

The word "orgasm," according to the *Oxford English Dictionary*, in older usage sometimes meant simply "a sudden movement or convulsion." An 1882 issue of the *Philosophical Transactions of the Royal Society of London* says, "Closely crowded together, the little party [of fish] ... suddenly would indulge in an orgasm; and lashing and plashing the water in all directions with their convulsive movements, would scatter at the same instant the eggs and sperm." But who knows what fish are really feeling, if anything? What is being described here could be a purely instinctive behavior on the part of fish during their release of eggs and sperm and has nothing to do with orgasm as we humans experience it.

Most animals seem to reproduce just fine without the pleasure of mammalian-style orgasm. Bedbugs, worms, mosquitoes, spiders, crabs, fish, frogs, and lizards perform their various sex acts with what appears to be an instinct, mindless and emotionless. The process is machine-like, a physiological reflex like a cough or a sneeze. In most animals, pleasure in copulation is not required for the continued survival of the species; copulation by itself is enough. Their nervous systems are hardwired to act on instinct—without the guide of emotions, without liking or disliking anything. An earthworm doesn't burrow into the soil and think, *how cozy*. A lizard doesn't contemplate the coming swarming season and think to itself with pleasurable anticipation, *how tasty those flying termites are going to be*.

Theoretically, ejaculation need not be accompanied by a pleasurable orgasm. The act could be a purely physiological response, as humdrum and unromantic as urination. But for mammals, the pleasure of copulation is something built into the nervous system, a kind of instinct, one in which desire and its satisfaction have taken over the more purely reflexive actions typical of non-mammals.

The pleasure of orgasm is not unique to humans. Its signs have also been observed in other animals. In chimpanzees and monkeys, at the end of a copulatory bout the male may show body tension and body tremor at the moment of ejaculation. The female sometimes exhibits muscle tremors and a "climax face" while experiencing rhythmic vaginal and uterine contractions. Both sexes have an increased heart rate and may emit rapid panting grunts or high-pitched squeals.

Even rats have orgasms, according to physiological and behavioral observations. During copulation both sexes emit squeaks not produced at any other time. The sexual responses of anesthetized rats stimulated electrically or by drugs, and of conscious rats freely moving about with miniature implanted transmitting devices, show that the physiological events of ejaculation in the male and the contractions of the pelvis, vagina and uterus in the female parallel what happens in humans. After a bout of sexual activity, the rats show signs of relaxation and exhaustion. They may later show a preference for the same partner and the same location in the cage, so the experience seems to be pleasurable for them. These findings are as close as we can get to answering the question of rat orgasms without actually being a rat.

Similar although more limited observations with similar results have been made on rabbits, dogs, pigs, cattle, and horses. What is true for chimpanzees, monkeys, and rats is probably true for all mammals: the sexual organs are connected to the pleasure centers and pleasure circuits in the brain. The linking of copulation with pleasure, generating pleasurable sex as opposed to purely mechanical sex, aside from any reproductive outcome, is a trait as old as the class Mammalia itself. The connection was probably made at least a hundred million years ago during the transition from ancient pre-mammals to ancient mammals.

That sexual activity in mammals is connected to pleasure is also apparent from other observations. Sexual activity is not always directly related to actually making offspring. Mounting of females by males has been observed outside the females' fertile periods in monkeys, antelopes, and goats. Same-sex sexual activity occurs sporadically in dolphins, elephants, monkeys, gibbons, bonobos and, of course, humans. Masturbation by both males and females has been witnessed in dozens of our primate relatives, from the lowly lemur to the lordly gorilla and many monkeys and apes in between. It has also been seen in dogs, cats, goats, horses, rhinoceroses, deer, camels, elephants, hyenas, dolphins, walruses, squirrels, and bats. What's all this non-reproductive sexual activity for, if not for the pure pleasure of it?

For the male, the wanting-and-liking brain provides a biologically straightforward connection between copulation and reproduction. Copulation leads to orgasm and the ejaculation of sperm cells. If the female is ovulating, fertilization of an egg is almost inevitable. A million years ago, accumulating gene mutations gradually led to the making of a new kind of nervous system, one that relied less on purely instinctive physiological reactions and more on memories of the pleasure of past copulations. This motivated the male to seek more of the events of body and brain that we call orgasm. The males' persistent search for orgasms led naturally to more offspring, who would also carry those modified genes.

For the human female the connection between orgasm and

reproduction is less obvious. Fertilization could occur perfectly well without female orgasm. Why is the female not just a passive vessel, who need take no particular pleasure in the procedure? As Marlene Zuk asked, "Why do men need the reinforcement of orgasm while women can reproduce perfectly well lying back and thinking of England?"

It may be that orgasm really has no real biological function in human females. Orgasm in the female could be just a useless by-product of very early development in the fetus when the first connections between the nervous system and the reproductive tracts are being made. Those connections for orgasm eventually become fully developed in the mature male, less so in the mature female. A parallel incomplete development happens in the male. The nipples of the mammary gland are formed early in the fetal stages of both sexes but eventually become fully developed only in the female.

On the other hand, female orgasm might have a biological function after all. Instead of adopting a take-it-or-leave-it attitude toward copulation, women in the long run might have more children if they show a statistical preference for "I'll take it." Why would they? There have been several suggestions:

- The female orgasm gives a boost to the sperm cells of a preferred man, sending them deeper into her reproductive tract where they have a better chance of fertilizing her egg than if she has no orgasm.
- The male partner who can provide orgasms for the female might in the long run somehow turn out to be a better father for the children.
- The risks of pregnancy and childbirth are potentially life-threatening. Without a pleasurable experience to offset those risks, a woman might tend to avoid intercourse altogether. To the extent that female orgasm is genetically determined, an orgasmic woman would on average tend to pass on her female pleasure genes to more children in the following generation, while the non-orgasmic woman, tending to avoid intercourse a little more often, would pass on her genes to fewer children.
- Female orgasm strengthens an emotional bond between a woman and a man. That would naturally lead to a longer-term association, which benefits the survival and well-being of the children.

The available evidence is not definitive. Sperm cells seem to make their way to the egg equally well with or without the woman's orgasm. There is no evidence that the best lovers make the best fathers. Neither a woman's fertility nor her instinct to bear children seem to be correlated with her having had an orgasm just prior to conception; indeed, many women don't have orgasms, but enjoy large families anyway. It's possible that emotional bonds between mother and father are strengthened by orgasms, but it's not clear whether this has any long-term effect in child-rearing. After all, female chimpanzees and female rats raise their offspring by themselves, and they have orgasms just the same.

Nevertheless, this aspect of the yin and yang of the coordinated evolution of males and females might hypothetically have arisen in the following way. Among our distant mammalian ancestors, males might have had orgasms as an enjoyable accompaniment to ejaculation and an inducement to seek out more females for the pleasure of it. An orgasm in the male might occasionally, almost accidentally, have induced one in a female, perhaps initially as a result of partially developed connections between her clitoris and vagina and her brain, a connection between genitals and brain that was more fully developed in the male. Her response might often have led her to favor him as sex partner. It also might have told *him* that she was likely to be available for sex again in the future. More sex between these two meant they would have produced more children between them, with more children carrying the genes for the genital-brain pleasure connection. In subsequent generations a stronger genetic link was gradually forged between pleasure and sex in both male and female.

It would be as if a man desirous of entering any door that opens to him is strolling along a block of row houses and discovers a woman who responds to the doorbell—and gets pleasure from what happens next. Eventually she comes to open the door more readily just for the sound of the doorbell, provided it is that same man who is ringing. Naturally that is the door to which the man will return, happy that the woman living in that particular house has learned to like the sound of the doorbell.

With this arrangement in place in the sexual behavior of our ancient ancestors, more offspring would be forthcoming—and with time the genes supporting such a system would come to characterize all the members of the species.

Ovulation Now?

In some mammals (rabbits, domestic cats, African lions) the release of eggs from the ovaries is a response to copulation itself. But in most female mammals, ovulation occurs on a regular schedule unrelated to copulation. The ovarian cycles include a period when the females are fertile and a longer period when they are not. In some species the fertile period is timed for a particular season of the year. In others it happens according to an internal clock—in the mouse, for a few hours every few days; in the wolf, for a couple of weeks every year; in the human and a number of monkeys, for about week every month.

In most mammalian species the female advertises when she's fertile. Depending on her species, she may do this by means of a visible change in her appearance, or special odors, or a change in her behavior, or some combination of these. The female tarsier, a small nocturnal primate of southeast Asia, marks her surroundings with vaginal secretions and urine to indicate her readiness for copulation. In the chimpanzee, the female's rear end swells up and takes on a rosy, pink color for about 10 days in the middle of her menstrual cycle, and that's when the males take special interest in her. The female gorilla, when she's ready, approaches the male, makes eye contact, reaches toward him and, if necessary, slaps the ground to get his attention.

Advertising makes sense. If you're a male and can read the signals, why waste sperm in unproductive copulations? For either sex, why spend time and energy in useless courting exercises? The female's anatomy, physiology, or behavior should say to the male, "Come on, I'm ready now." At other times she should indicate, "I'm not ready at the moment; don't bother me."

There are approximately 500 primate species in the world. In most of them the female advertises her fertile period in one way or another. Given such a wide distribution across primates, it seems likely that for primates it was the ancestral condition to advertise. There's only a small handful of primate species in which the females *don't* overtly advertise their fertile period. They include the wooly spider monkey ("muriqui") of Brazil, the Assam macaque of southeast Asia, the vervet monkey of southern Africa— and humans. In these species the trait of advertising the fertile period was lost for some reason.

Was there no longer any advantage in females' advertising the times when their eggs were available for fertilizing—no advantage in efficiency of mating, in effort saved, or in number of offspring? Was the advertising relinquished as not worth the physiological or physical effort? "Fertile signs *abandoned*" would be the appropriate phrase to describe this evolutionary development.

Or might the actual *suppression* of the ancestral signs of the female's fertile period somehow have had a reproductive payoff? The phrase "fertile signs *concealed*" would then be the right phrase. The expression commonly used for humans and the few other primates that exhibit it is "concealed ovulation," and that seems to be the correct evolutionary conclusion.

In humans and in the other primates where there are enough observations, the lack of obvious signs of when females are in their fertile period in their ovulatory cycle is correlated with extended sexuality—females and males mating at almost any time in the cycle. That's true for humans, for Assam macaques, and for wooly spider monkeys. In humans, it's also associated with other permanent attractions for the male: the enlarged breasts and buttocks of the sexually mature female, which do not cycle with fertility and lactation but remain as signs of sexual maturity. The evolutionary signal is that the female is ready for copulation—other conditions being met (the right social circumstances and the suitability of the male) at any time in her cycle.

How might this state of affairs have arisen, given ancestors in which the females advertised? It may be that the human female evolved to be sexually attractive to males all the time, and not just during her fertile period, because she might then have had a greater range of quality males to choose from to father her offspring, not just those who happened to be around at the right time. This might mean better fathers and more and better offspring. So new genes that resulted in the suppression of the female's signs of ovulation would have gradually been distributed to greater numbers of offspring as the generations went by.

It is also quite possible, even likely, that in humans, sexual activity evolved to mean more than just reproduction. It seems to have come to be used for routine pleasure and long-term bonding. (That *homo*sexual activity is as common among humans as it is, more common among humans than in most other primates—except bonobos, who also use sex for more than just reproduction—supports the idea that with us sex is more than reproduction.) Babies often result from the continual heterosexual activity, but that is almost a side effect. It is to be sure a necessary one for the continuation of the species, but not our main motivation for engaging in it. So it is natural that the females' overt advertising of their fertile periods would gradually diminish over evolutionary time. Periodic signs of the fertile period might have been abandoned as being beside the point.

But signs of the human female's fertile periods may have been not simply abandoned over evolutionary time but *extinguished*. This could have happened if greater numbers of healthy offspring *did* result, even if she didn't show off outside what her hormones were doing inside—at least not so that most males would notice and even if a majority of her copulations occurred outside her fertile period. An ancestral female genetically predisposed to concealing signs of her ovulations might have had *more* offspring, and healthier ones, and might therefore have passed on more of those genes to those offspring than she would have if she advertised her fertile period. How might this have happened?

With the man continuously interested and the woman continuously receptive, the extra copulations may have resulted in more children because they were taking advantage of the beginning and the tail ends of the woman's variable fertile periods, even if she was not advertising them. On average they might have had a few more children than if the man was paying attention only to the more blatant female signals. The concealment of ovulation may have turned out to be a good thing after all, because it led to more male attentiveness and therefore inevitably to a greater number of children.

The reproductive output of both the man and the woman may also have increased as a result of the pleasurable closeness they both experienced from the extra copulations, even the non-productive ones. Both parents may have stayed together longer during the time when the children were growing up. The children would have reaped the benefits of having two caring parents—the one who provided the milk and day-to-day care and the other who provided the protection and extra sustenance for the family. Because of the greater number of surviving children, the genes supporting those parental instincts would have been represented in ever-increasing numbers in succeeding generations.

This picture of the evolution of the human mating system has the appeal of fitting the human concealed-ovulation trait into a mosaic of other traits that were evolving in concert, all fitting together like pieces of a jigsaw puzzle—orgasm in females, seemingly unnecessary; the years of care the human child requires because of its helplessness; and the likelihood of there being several helpless infants and children to care for at the same time, a task that would be difficult for the mother to carry out on her own.

The closeness generated by mutually satisfying sex outside the female's fertile period may also have its own benefits, unconnected with producing offspring. The male and female are not just potential parents but also a social team providing mutual aid and comfort. In the Assam macaque the female does not advertise her fertile period, is sexually receptive throughout her ovarian cycle, and concentrates her copulations on a preferred male. So with the Assam macaque, having sex seems to serve a social as well as a reproductive function. We humans can relate to that.

The series of evolutionary events that led to our mating system can be summarized as follows. A few million years ago ecological changes in Africa led to the evolution of an ape species that was two-footed, walking and running upright. Later ancestors on that evolutionary line also evolved larger brains. That made them cleverer and gave them new skills in tool-making, foraging, hunting, cooperation, and communication—a whole host of advantages for survival and increased efficiency in producing offspring. Meanwhile, to support the upright body, the pelvis changed and the birth canal narrowed at the same time that evolution was producing a larger head for the newborn. These conflicting requirements created the "obstetrical dilemma"—big head, narrow birth canal—for the newly emerging human species. The evolutionary compromise was to have the newborn come into the world before its brain was fully developed, before it could begin to care for itself. The human mother, unlike the chimpanzee or gorilla or orangutan mother, would have had difficulty raising her offspring successfully by herself, especially if another newborn appeared within a year or two and maybe another one after that. Further evolutionary adaptations followed: on the female side: extended sexuality and gradual suppression of signs of her fertile period and, on the male side, an expansion of his female-oriented sexual attentions to longer and longer periods before and after her ovulation times. There were two automatic side effects. His more continuous presence provided extra protection for the female and her offspring—who, with increasing likelihood, were also his. His constant attentiveness included bringing back more antelope meat from the hunt to be shared with her and her children. New forms of genes modified his behavior toward increased protection and provisioning of his female and the offspring. With greater offspring survival, her "ovulation-concealing" and "ready-for-sex-at-any-time" genes and his "more-constant-attention" genes were passed on in greater numbers to the next generation, and eventually came to characterize the whole species.

The newly evolving human followed its own evolutionary path, unlike the ones taken by the ancestors of modern chimpanzees, gorillas, and orangutans. For them, maternal care by itself was enough for the offspring because of their smaller brains, easier births, earlier independence of young, and the easier requirement of usually having to care for just one immature ape at a time. We humans evolved our own route to reproductive success and that has made all the difference.

Monogamous or Polygamous?

As far as our mating arrangements are concerned, what is our true nature? Did our evolutionary ancestors follow a rule of lifetime monogamy, one male with one female? Or were they polygamous—one male with several females, or one female with several males—or just generally promiscuous? Whatever they were, are we modern humans genetically constrained to behave as they did?

What do the other great apes do? If they all had the same sort of mating system, we might guess that our human ancestors had that mating system too, since we all came from the same evolutionary stock. Alas, the other great apes fail to provide us with a consistent answer. Chimpanzees choose one partner or another more or less indiscriminately and are quite promiscuous from one mating bout to the next. Gorillas have a harem system; the dominant silverback male guards and mates with several females, fighting off would-be male intruders when necessary. Orangutans are solitary most of the time, seeking out partners only for mating purposes. Among the orangutans of Borneo, a mating association between a male and a female may be only a matter of a few hours; among the orangutans of Sumatra, it may last for weeks. There is no evidence that the next time an orangutan mating occurs it will be with the same partner.

Surveys of a wider range of primates also fail to provide a consistent picture. Overall, monkeys and apes tend to be more often polygamous or promiscuous than monogamous. The hamadryas baboon has a harem system, with small groups made up of one male and his harem of several females. The olive baboon is more promiscuous, males sometimes guarding a particular female for copulations, but many females often copulating with several other males. Most gibbon species are monogamous, even more so than humans, and only occasionally stray outside their one-on-one relationships.

You might think it would be impossible to know anything about the mating habits of long-extinct apes, including our evolutionary ancestors. But there are clues from the patterns of mating systems of living primates. Aside from humans, the other great apes, both polygamous and promiscuous, have males that are considerably larger than the females. This is a result, it is thought, of natural selection for large male body size when the male of the species has to fight off other males for access to, or defense of, the female or females. The large size of males would initially have evolved because of natural genetic variation in body size. In the evolution of gorillas, say, the larger ancestral males would have won more fights and therefore would have passed on their genes to more offspring. Over time, gorilla populations would come to have larger and larger males, eventually reaching their modern condition. (Why didn't gorillas continue to evolve in that direction, giving King Kong–sized males now? At a certain size, it becomes hard to find enough food to feed an excessively large body, and an excessively large size, given the physiology of the animal, itself begins to cause health problems. As with many traits, evolution strikes a balance between too little and too much.)

As it turns out, orangutan males weigh twice as much as their females. Gorilla males weigh about 75 percent more than their females. Chimpanzee males are about 40 percent heavier than chimpanzee females. By contrast, the human male weighs on average only about 20 percent more than the human female, while the gibbon male weighs only about 10 percent more than his female. In gibbons and in humans, monogamy appears to have evolved as an alternative means of maintaining productive male-female pairings without physical combat, and male-female size differences are less.

Correlated with the larger male body size in species in which males fight for females are the males' large canine teeth, weapons used in the fights. In the monogamous gibbons and in humans the canines of males are not noticeably larger than those of females.

Therefore, where fossils of extinct species are numerous enough and complete enough, the evidence from teeth and long bones allows something of their mating habits to be inferred. On this basis, it looks as if *Ardipithecus* ("Ardi"), *Australopithecus* ("Lucy"), *Homo ergaster*, and *Homo heidelbergensis* (see Figure 23) all tended toward monogamy. Monogamy, or at least some system that avoided physical conflict among males, appears to have been the usual practice of our evolutionary ancestors from millions of years ago. Such mating systems are likely ingrained in our genetic makeup. The evolutionary progression may have run as described in the previous section: upright walking, narrower female pelvis, large brain of the fetus, helpless infants, sexual arrangements that kept the male attracted to and tied to the one female (i.e., monogamy), and consequently better provisioning and better survival rates of the offspring.

Yet while our genes may predispose us toward monogamy, ours is not a completely fixed, genetically determined mating system. We are a variable species, flexible in how we deal with our reproductive practices. Surveys of hunter-gatherer societies the world over show that humans have different arrangements for provisioning and caring for a woman's several dependent offspring at a time. Some societies have the common system of children being protected and cared for by one mother together with one father. Others have various amounts of additional help—from the father's relatives, from the mother's relatives, from the several men who have had sex with the woman prior to and during her pregnancy, or from the community at large.

Variety in how the biological mother and father mate and raise their offspring is also known in other primate species. The common marmoset of Brazil lives in family groups of a dozen or so individuals, typically dominated by one breeding male and one breeding female and their offspring. Their mating system is "loosely monogamous"—the dominant male and female occasionally copulate with others both within and outside the immediate family group. Atypically for primates, marmoset births are usually twins and occasionally even triplets or quadruplets. This would place an extra burden on the mother, but the father and other adult members of the family group cooperate in taking care of the juvenile members.

Mating system variability is also seen in the "swamp monkey" of western Africa. It is officially named "de Brazza's monkey" after the 18th-century Italian-French explorer Pierre Paul François Camille Savorgnan de Brazza, who first described the species. Singlet births are the rule, and care and feeding of the young falls exclusively to the mother. In western parts of its range, populations consist of small family groups, one faithful male-female pair and their offspring. In eastern regions, one dominant male lives with several females and breeds with all of them. The reason for the difference in mating styles is not known but may have something to do with adjustments to different ecological circumstances. Who knows?—the same may be true for humans. The point is that different breeding and family arrangements are possible within a species.

Primate breeding systems are therefore highly variable from one species to another and even within a single species. Humans worldwide are similarly malleable in their mating patterns: sometimes strictly monogamous, frequently serially monogamous (one wife or husband after another), occasionally promiscuous, and harem-holding when laws and customs and male finances permit.

Menopause

Animals everywhere come into the world, reproduce, and then die soon after. A mayfly hatches, lives a day or two, mates and lays its eggs, and then dies. The California two-spot octopus female matures in a year or two, lays a single large clutch of fertilized eggs, and then dies. Pacific salmon, after a few years in the ocean, return to their natal streams, release their eggs and sperm, and then die within a few days. A female mouse is born, matures in a month, and reproduces continuously over the next couple of years, and then dies. A female wolf may continue to produce litters every year until shortly before her death. A female chimpanzee is fertile into her mid–40s, usually just a couple of years before dying. A female elephant produces a new calf about every six years until her old age, around 60, and then she dies. All this makes sense if the biological purpose of life is not to live a long and happy life but to produce offspring.

One of the evolutionary mysteries of the human mating system is menopause. The young human female reaches sexual maturity in a little more than a decade, traverses her reproductive phase of three decades or so—and then shuts down her reproductive system. But contrary to the general rule, and not the typical pattern for animals, she lives on for many more years. Evolution generally rides along on having as many offspring as possible over a lifetime. Why stop having children when you have many more years to go?

Menopause happens to women but not to men. It makes evolutionary sense for men to be able to father children well into their 60s and 70s. As long as the male can continue to provide adequate support for the resulting offspring, or if he is needed for little else but to provide sperm cells, then genes supporting the aging male's ability to continue to sire children would automatically result in a greater number of children, those bearers of just those "sperm-in-old-age" genes. After providing sperm cells, the man has relatively little more to do in making offspring; the burden is mostly on the woman. She expends most of her reproductive energy in the development of the fetus and in making milk for the newborn child. The explanation for menopause must lie in how she spends her effort in long-range reproduction.

It can't be just a question of modern medicine's providing an artificial extension of the female lifespan. Even in the 17th, 18th, and 19th centuries, women often lived into their 70s but still stopped having children in their 40s. In isolated African and South American tribes, where people don't have access to modern medical care, women can live well into their 60s; they still undergo menopause at the usual time. So it's not a question of the modern woman's extending her life beyond a normal reproductive span. It's a question of the woman's mid-life self-sterilization. Our species has "female early sterility genes" that cut the female reproductive life short.

There must be some evolutionary advantage for our species to have its females cease producing offspring in their advanced years, in contrast to the way of chimpanzees and elephants and most other animals. What could that advantage be, and why would it apply to humans and not to other animals?

Some of the possible explanations are these:

- The reproductive systems of older women shut down from lack of use because men prefer younger women as mates.
- Men tend to leave their women after a while, and there's no genetic point in a woman's maintaining her reproductive ability if there's no man around to help raise the kids.
- Older women's bodies withdraw from the evolutionary fray because they're unable to compete with their younger and more fertile daughters, who hoard all the available resources for themselves.
- The reproductive lives of women are cut short as they age because of the drain on their own physical resources, and because with an extended lifespan and less effort they can help care for nieces, nephews, and grandchildren. This is a way of indirectly extending their reproductive lives, caring for the next generation or two, not just their own children.

There is biological precedent for self-sterilization. The question arises starkly in the case of other social animals where many females are sterile from the moment they come into the world. The classic example, to which Charles Darwin himself gave puzzled attention, are the sterile workers of the honeybee hive. These daughters of the queen create a hive that supports the queen's efforts in producing a few fertile daughters. The queen produces more queens of future hives by having a sterile workforce than she could without it. The workers themselves, of course, can never pass on their sterility-causing genes. But they help their queen do so. The new queens produced in the hive can fly off and, in their turn, make their own sterile workers, and thrive as a result. The sterility of honeybee workers in the hive is all about indirectly helping the reproduction of their mother (the queen) and their sisters (future queens).

In human terms, late sterility—menopause—might be a good thing for the sake of earlier children, especially given the extensive, multi-year care that human children require. If human mothers lived another decade or two after 30 childbearing years, they might on average have more healthy offspring by concentrating their energies on their existing children rather than on new offspring produced late in life. Imagine raising four children successfully, while on the other hand trying to raise six might put your family into debt or impair your own health. Imagine undergoing six pregnancies in a row starting at age 16. At least some of your children would be likely to suffer from inadequate care. (As a woman friend of mine said, "After a certain age you just don't have the stamina, patience and all else that goes with raising a young child that you have when you are younger.") In the long run you might leave more descendants by raising four healthy kids rather than six sickly ones. A "menopause gene" might foster that better outcome. The number of copies of such a gene would increase if healthy offspring rather than sickly ones were the bearers of that gene in succeeding generations. Like copies of letters in envelopes bearing the proper postage rather than in ones with inadequate postage, more gene copies are propagated if they are safely housed in a small number of healthy bodies than in a large number of unhealthy ones. If the letters themselves (the genes) were able to copy themselves with the right amount of postage on their envelopes (the bodies), after a while you would find only letters in envelopes bearing the right postage.

A woman might also have more descendants by caring not only for fewer healthy children, but for her grandchildren as well. This is the "grandmother hypothesis." Second only to a mother's love for her children is a grandmother's love for her grandchildren, born and unborn. How often does a mother say to her married daughter, "When are you going to get pregnant?" And young parents who look after their own aging parents could indirectly be benefiting their own children by being able to enlist their own parents' experience and help. Given the difficulty of raising human children with their long years of dependency, the more help the better.

You, as a woman carrying most of the child-rearing effort, might be able on your own to raise successfully, say, four children. Theoretically each of them could then raise four children of their own, giving you 16 grandchildren. But if with the help of your own mother you could raise *five* healthy children, and each of them could, with *your* help, raise five healthy children of their own, then you would have 25 healthy grandchildren instead of only 16. What counts evolutionarily, in the long run, is not only the number of children you produce, but also the number of grandchildren (and the number of great-grandchildren, and so on).

15

On Being Social

"…contemplating the first and most simple operations of the human soul, I think I can perceive in it two principles prior to reason, one of them deeply interesting us in our own welfare and preservation, and the other exciting a natural repugnance at seeing any other sensible being, and particularly any of our own species, suffer pain or death."
—Jean-Jacques Rousseau, *Discourse on the Origin and Basis of Inequality Among Men* (1754)

As a part of normal development, one forcefully asserts his or her independence at least twice in life, first around two years of age (the terrible twos), a stage not preserved in memory, the one where the child shouts "No!" and throws a temper tantrum when its demands are not met. The second time occurs in early adolescence, usually around 13 to 15 years of age. I remember telling my parents, in my mind, silently, "Your values are not my values; I'm going to make my own way in life and find a Truth you seem to know nothing of." This kind of sentiment is the first step toward becoming a mature adult. We then go out into the world, independent of social ties, a self-made man or woman.

This is an illusion. Everywhere we are bound by social ties, slender invisible threads like steel-strong spiderwebs tying us in countless ways to other members of our species. Without them we would starve to death, we wouldn't learn to speak, we wouldn't learn to suppress our purely selfish instincts when the social circumstances demand it. Our untethered behavior would raise the hair on the backs of the necks of other members of our species. As we grow up in society. there are also the day-to-day mental and emotional connections that we naturally forge with others, developing values and morals, building a sense of honor and purpose. Whether we will it or not, our brain develops to make those social connections, inside and out: brain wiring inside, and the resulting sensitivity, outside, to the minds and behavior of others.

What does it mean to be *social*? It's not the same as being *sociable*. When Veronica snubbed Betty at last week's dinner party and then insulted her again when next they met, Veronica was not being very sociable. But she was being highly social. She was sending a specific message to another member of her species. She intended to influence Betty's subsequent behavior toward herself, as well as others' attitudes toward Betty.

The dinner party itself was social behavior exemplified in all its glory: ritualized communal eating, division of social roles according to hierarchical seating arrangements, division of labor into kitchen staff and waiters, not to mention the army of farmers and vintners and manufacturers and builders and transportation workers who raised the domestic animals, grew the grapes, constructed the table and chairs, provided the dinner plates, soup bowls, utensils, tablecloth, candlesticks and wine glasses; also those who designed and built the house where the dinner party was held, those who drove the trucks that transported the food and wine to the table, made the apparel and the shoes of the dinner guests, and the guests' tailors, barbers, hairdressers, and manicurists.

All this division of labor and interdependence may merely be the trappings of modern culture and society, and not fundamental to the human condition. It is nevertheless an elaboration of human social life that already existed in the earliest members of our species. Hunter-gatherer societies (examples of which are the small and relatively still-isolated groups of people living in central Africa, central Australia, central South America, and the Arctic regions of North America, Europe and Asia) may be our best models of how humans might have lived before the rise of agriculture and civilized society. In hunter-gatherer societies, different tasks—the gathering of food, fuel and water, making clothes, hunting and fishing, constructing and maintaining shelters, constructing boats and nets, making tools, and making trade agreements with other groups—are typically relegated to different people, often along gender lines. Individuals in these societies also have their own specialties, one person expert in hunting, another in repairing tools, another in weaving, another in knowing where the best fruit can be found, and so on. And everyone in the group benefits from sharing the fruits of the labors of the individual members.

Social ostracism, solitary confinement, and facial nerve paralysis (where the individual can't produce the normal facial expressions that go along with emotions) wouldn't be the traumatic experiences they are if we weren't inherently highly social creatures, needing to communicate with our own kind on a regular basis. Why else are there dinner parties?

In addition to snubs and insults, dinner parties, and mutual interdependence of all kinds, other signs of human beings' complex sociality are the social interactions such as helping and seeking help, cooperating, cheating, stealing, negotiating, imitating, teaching, learning, contracting, buying, selling, borrowing, lending, taking charge, taking offense, taking revenge. It's all sociality, if not always sociability.

A measure of how completely social a creature the human being is comes from a consideration of this question: *in what ways might a human act as totally independent being?* Humans are in fact similar to ants, termites, and honeybees, whose individuals have no life whatsoever outside of the colony. Even St. Jerome throughout his time as a hermit in the Syrian desert continued his theological writings and studies. Basho in his wanderings throughout Japan depended on strangers and friends for food and shelter and conversation, as did the mendicant monks in Europe in the Middle Ages. Jeanne Le Ber, living as a recluse in 18th-century Canada and vowing a life of seclusion, poverty and chastity, nevertheless retained title to her family property, went to mass, had her cousin attend to her personal needs, sewed clothing for the needy, willed her estate to a convent, and asked to be buried on Church grounds. Arthur Leslie Darwin, the Hermit of Opossum Key, lived without electricity, caught rainwater in a cistern, grew his own vegetables and raised his own rabbits. But he built his shelter out of concrete blocks manufactured elsewhere, owned a battery-powered radio, and every few weeks went into Everglades City for supplies.

The selfish new-car buyer and the self-serving thief are completely embedded in an already-established economic system. The lonely artist in his garret and the poet in her isolated farmhouse are creating their paintings or poetry not in total isolation, but for an imagined posterity.

Basic Elements of Being Social

Social behavior, or sociality, comes in many degrees and many forms. There are the loose associations exemplified by schools of fish and flocks of birds. There are the more tightly organized colonies of honeybees, ants, and termites, where different tasks are divided among different castes. In all of these insects the individual doesn't count for much. That kind of sociality has its counterpart in some kinds of human association—patriotic parades, party conventions, protest rallies, theater audiences, religions, workers' unions, and identities such as "chemist" or "conservationist" or "Manchester United fan" or "native of France."

Human sociality can be defined and categorized at different levels. There are individual level, one-on-one personal interactions; relations of one group to another; and society-wide associations that encompass politics, religion, economics, international relations, and history. Someday we might be able to integrate the different levels in one overall explanatory scheme, just as one tries to understand the coordinated behavior of a school of fish in terms of the interactions of individual fish.

Human social complexity can be illustrated by the life of our fictional Betty, mentioned above.

Betty went to school, graduated near the top of her class, got a law degree, took a year off to work in the Peace Corps in Guatemala, and then returned and passed her bar exam and obtained a junior associate's position in a New York law firm. She now works alongside others specializing in patent issues, consumer safety affairs, and tax law. She hopes to work her way up to senior associate and maybe eventually full partner in the firm. She does *pro bono* work for the Center for Justice and Accountability. She donates to the Save the Children Federation, Human Rights Watch, the American Cancer Society and the Jazz Education Network. Once a week she goes to an early morning yoga class. Once a month she attends a book club. On Fridays she gets together with friends downtown at McSorley's Old Ale House. Sitting at one of those old wooden tables, she loves soaking up the place's literary history; e.e. cummings was a regular there. He wrote, "I was sitting in mcsorley's. outside it was New York and beautifully snowing. Inside snug and evil ... the break on ceiling-flatness the Bar.tinking luscious jigs dint of ripe silver with warmlyish wetflat splurging smells waltz the glush of squirting taps plus slush of foam knocked off...." Even though she's just a reader, she still likes to think that she's part of an ageless literary tradition.

Like many other primates, human beings naturally organize themselves into hierarchies. Betty and her office mates all work on consumer safety issues; at the same time, they have a sense of belonging to the legal department of the firm, as distinct from the financial, administrative and information technology departments. Beyond that, Betty also feels an allegiance to the whole firm. She's also aware of herself as a New Yorker, and when she travels abroad, she's an American. She has no difficulty keeping all her allegiances straight.

The human brain seems to be wired in such a way as to trigger different intensities of feeling of belonging, depending on the size of the group. The brain may have evolved to its present large size, in part, in order to be able to deal with the increased mental effort required in keeping track of the ever-increasing complexity of social interactions. A friend of mine put it this way: "In general, I find groups of people to be confusing. One on one I can listen closely enough and given enough time I can have a sense of what's going on with the other person and what's going on with me. But a group is a whole different beast, with far too many things going on for me to make much sense of what is in its heart or in the hearts of all the individuals within the group. In work settings, in that kind of group, I've been aware of hierarchy, and that sometimes people are jockeying for power, and that sometimes there is a subtext of some sort that I find difficult to read, let alone how to be in that conversation, or even whether I want to be in that kind of conversation."

A feeling of loyalty goes out first to small groups, such as family or a small circle of close friends one sees daily; secondarily to larger groups such as work colleagues, individuals among whom one comes into direct contact less often; and finally, more distantly, to larger groups that are more abstractly defined, such as the school where one was educated, the organization in which one is employed, or the city or state where one lives, most members of which one hardly ever, or even never, sees. The opposite side of the loyalty coin is the depreciation of "the other"—any member of a group that's different from and maybe antagonistic to one's own. Such tribalism is woven into the very fabric of the human mind; it undoubtedly has evolutionary origins. The ancient communities were able to thrive not only because of the loyalty of their members, but also because of their automatic, unthinking antipathy toward other groups. This is the basis of racism and other forms of reflex prejudice and stereotyping.

Human beings, uniquely, form large, organized assemblies that perform or watch other members of the species carry out rituals and demonstrations of various kinds—in churches, mosques, synagogues, football and baseball stadia, classrooms, theaters, concert halls, marriage ceremonies, funerals, political rallies, music festivals, and parades. Into this category of spectator events fall the Metropolitan Opera performances that Betty attends and the performances at the Blue Note Jazz Club she goes to on Saturday evenings. Social as baboons and chimpanzees are, you don't see them organizing events and sitting in rows to watch other members of their species perform.

Betty is a member of a nation state, the middle class, a representative democracy, Western culture, an industrial society, and urban sprawl. These also describe aspects of human sociality, although Betty is not ordinarily conscious of them.

Betty's life illustrates the basic elements that go into the making of a human being as a social creature, the one-on-one interactions that are the basis of our wider sociality:

- fitting into hierarchical organizations and conforming to social norms;
- learning by watching and imitating, and being receptive to teaching;
- cooperating with others in group activities and endeavors;
- feeling empathy and compassion, and behaving altruistically; and

- being able to read, automatically and without conscious analysis, the beliefs, intentions and desires of others.

These traits are also exhibited to some degree by some non-human animals. In humans they are amplified, making us one of the most social of animals.

Teaching and Learning

There are two ways to learn how to navigate life's vicissitudes and to survive and flourish. One way is to learn by trial and error—always somewhat risky, because a trial followed by a serious error may do you in. "Learning from experience" is a long-term, slow-motion trial-and-error method, if you are lucky enough to survive your experiences. For aardvarks, platypuses, skunks and other solitary animals, experience with the world is all they have except for their instincts, built-in, genetically determined behavior patterns. These are useful and efficient where the environment is more or less predictable.

You have a second option if you're at least a somewhat social animal: social learning, learning from other members of your species. You can learn from your parents or from others in your group by observing and imitating their behavior. You can learn to associate their communications and their behavior with food sources or with danger, or by simply accommodating to the group, following your herd or its leader wherever it goes.

Birds learn their songs by listening to their fathers' songs. Rats learn how to find hidden food by watching other rats. Young dolphins learn to communicate with one another by practicing their vocalizations, and they can learn practical methods of searching for food by observing their mothers. Young elephants don't have to search out waterholes on their own; they learn the locations by following the herd. Young chimpanzees copy their mothers in learning how to use stones to crack open nuts.

Clear instances of teaching of naïve young individuals by experienced adults are rare in the animal kingdom outside the human species. There are a few examples. A mother cheetah will sometimes refrain from killing the rabbit or small antelope she has just caught; instead, she'll take it back for the cubs to practice on. Meerkats, the small mongoose-like animals of southern Africa, whose diet includes scorpions, do something similar. The meerkat mother will bring a live scorpion back to the burrow, remove its sting, and allow her pups to play with it before eating it. From the experience the young meerkat learns how to handle scorpions on its own.

The vervet monkey of southern and eastern Africa has a system of communal alarm calls that tell from which direction a particular predator might be approaching. A short series of harsh squawks or barks signals a leopard sneaking through the bushes. A croak-croak-croak call means an eagle is swooping down from above. A sequence of chattering sounds means a python is slithering through the grass. If one monkey spots the danger, others in the group check it out and join in the chorus. Very young monkeys learn the system, getting better at it over time. They learn partly by simply imitating the adults and also by being an adult encouraged by other adults. When a young monkey is the first to spot danger and give the right call, adults may join in with the same call, reinforcing what the young one is communicating. If a young monkey gives the wrong kind of alarm call, its mother (after first fleeing) may return and give the youngster a slap or a bite. This kind of interaction is observed often. It certainly seems that vervet monkeys are in fact teaching their young, human-style.

In the coastal forests of southeast Brazil there lives the golden lion tamarin, a beautiful little monkey with long golden fur. It eats fruit and nectar from flowers, as well as beetle grubs, spiders, and caterpillars, which it searches out under bark and in wood crevices, vine tangles, rotten logs, and leaf litter. Adults have a special "here's food" call, which they use for the very young. Later the adults use the same call for juveniles who are old enough to begin foraging for themselves, showing them the kinds of places where food might be found. The juveniles then dig out the insects and spiders on their own. In this way the adults teach the young the best kinds of places to find food. Give a young monkey a few grubs, and you feed him for a day; teach him where to look, and you feed him for a lifetime.

Teaching-and-learning practice must have been built into the primate brain from the beginning tens of million years ago.

Early humans advanced teaching practices even further. In today's hunter-gatherer cultures, as it must have been in earlier humans, fathers teach their sons how to make a spear, how to hunt, how to skin the antelope. Boys practice their skills in make-believe play. Mothers teach their daughters where to find the best wild plants, how to use a machete, how to cook, how to weave clothing, how to care for infants. This kind of teaching goes beyond what other primates do. Adults also teach young who are not their own offspring. Adults teach other adults. The whole process of teaching is facilitated by language, itself a skill that is shaped by teaching. Human primates sitting around the campfire, after the day is done, also teach each other by telling stories.

From such beginnings come trade apprenticeships, schools, colleges, and universities. Innovations that arise in the minds of creative individuals are eventually spread throughout the culture and are passed on to succeeding generations. Humans progressed step by step from first imagining that infectious disease might be the result of inhaling bad swamp air, to the discovery of microorganisms, to the germ theory of disease, to the development of antibiotics. We went from studying the flight of birds to building fixed-wing gliders, to manufacturing small gasoline-propeller-powered airplanes, to passenger airliners, to jetliners, to rockets for spaceflight and space exploration. Humans pile discovery upon discovery, innovation upon innovation, all without any further genetic changes in our psyche. Cultural change surfs forward on self-generated waves.

Cooperation

Working together toward some end for the benefit of all the participants is characteristic of individual human beings. It's also characteristic of corporations and their divisions and departments, of government agencies, and of whole governments and whole nations. At each level, entities cooperate with one another to accomplish a goal.

A good example of people helping other people is barn-raising (Figure 55), which was more common in the 18th and 19th centuries than it is now, although it's still practiced in some rural districts. Members of the community pitch in and help a farmer build his barn. It might seem that the average community member gets no direct benefit from doing this. It's not *his* barn that's being raised, and his time might be more profitably spent tilling his own fields or slopping his own pigs. But a day may come when he, or maybe his son, needs a new barn, and constructing one is something he can't do on his own. So having helped with another's barn raising, he can expect some help in return.

Members of a community may pitch in to build a bridge across a river, or sandbag its banks when the river rises, and everyone benefits.

Similar examples of cooperation in pre-agricultural times can readily be imagined—cooperative hunting of large game, where a single hunter by himself is unlikely to be successful; defending a village against a raiding tribe, where a single defender will almost certainly be killed; and foraging for ripe fruit in the forest, where a single pair of eyes is less likely to find any. These might be the forerunners of the more modern practices of cooperation in civilized society, such as barn raisings, community-watch programs, rowing crews, basketball teams, labor unions, political parties, the sharing of information between the Bureau of Economics and the Consumer Protection Bureau, the forging of trade agreements between nations. People cooperate with others when they build a house, play in an orchestra, sail a ship, design an airplane, or make a movie. Aided by language and often involving organizational hierarchies and a division of labor, cooperation is a hallmark of human social activity.

Figure 55. Barn-raising, a cooperative behavior (photograph from southeast Missouri by Russell Lee, 1938, U.S. Farm Security Administration, Library of Congress).

Cooperative behavior seems to rest on rational decision-making; we cooperate because it makes sense to do so. But there's a gene-based interpretation: if the cooperative spirit is undergirded by "cooperation genes" in the farmers, then the cooperating farmers will have larger numbers of children than solitary farmers, and the number of copies of those cooperation genes will increase as one generation succeeds another. Does this explain human cooperation? Could human cooperative behavior be dictated by genes that were passed down from early evolutionary ancestors?

Many animals also show cooperative behavior, even though they have never been accused of being rational actors. Bees, wasps, ants and termites cooperate in building their hives, nests and mounds. The sociable weaver birds of Africa cooperate in building their communal multiple-unit apartment-style nests. In a species of Australian skink, family members cooperate in constructing and maintaining an elaborate underground tunnel system where they all live. In all these cases you, the cooperating individual, are taking advantage of the activities of others for your own benefit—and all the while *they* are taking advantage of *your* activity to benefit themselves. The essential feature of cooperation is that the job gets done, to everyone's benefit—a job that could not be done otherwise, or at least not as efficiently or as well as if the job were to be attempted by an individual working alone.

Bees or ants or skinks don't reason things out. Each cooperating in its own way is just what these animals do. It's as integral a part of their nature as their anatomy. Underneath their cooperative behavior lies pure instinct, behavior dictated by genes. The evolutionary explanation is that in the distant ancestors of all these animals there were genes that in one way or another tilted behavior in the direction of cooperation. The individual in the group gave up little and in the long run gained much by cooperating. It works best if the members of the group all carry copies of the same "cooperation genes." This would be likely if they were all members of the same family, all genetically related. This is in fact the case for many of the animals that exhibit cooperative behavior. It's not true of all—humans, for example—so there must be evolutionary pathways by which cooperative behavior extended beyond the boundaries of the family to all the members of a population. Those distant ancestors, their offspring, and neighboring colonies and tribes served as containers which over time, in greater and greater numbers, came to carry those genes. Today they are the social insects, the weaver birds, the cooperative skinks, and the many other animals that behave in cooperative ways—human beings, too.

The social mammals, perhaps because of their greater cognitive powers, appear to have something beyond instinct. They seem to possess some extra ingredient governing their behavior, an awareness of what others are doing, or intend to do. They often adjust their behavior to fit in with others in order to attain a desired objective. Behavior that looks very much like that kind of cooperation has been observed in wolves, hyenas, lions, elephants, monkeys, chimpanzees and dolphins. Here are some examples.

Wolves, hyenas and lions sometimes hunt alone, but when the prey is a bison, zebra, wildebeest or buffalo, they usually hunt in groups in which the individuals coordinate their behavior. They fan out and surround the prey; some of them separate the prey from the herd while others attack it.

Bottlenose dolphins in a large tank or enclosed lagoon quickly learn to cooperate with one another in various kinds of tasks for the sake of a fish reward. Out in the ocean, male bottlenose dolphins cooperate by forming alliances for protection and for chasing and sequestering potential female mates.

Elephants in the face of a threat such as a lion on the prowl will cluster together, placing the young elephant calves in the center of the group for protection. Adult elephants cooperate in pulling an elephant calf out of a mudhole.

Cotton-top tamarins, the small squirrel-sized monkeys in the forests of Colombia, South America, live in small family groups. They occasionally include outsiders as well. All the adults, whether the actual parents or not, cooperate in taking care of the young, carrying them around and feeding them.

Chimpanzees in the wild occasionally supplement their usual vegetarian diet with the colobus monkey meat. Male chimps in a party of a half dozen or more will go out hunting and kill all the monkeys they can chase down. If a single chimpanzee runs across a group of colobus monkeys, rather than chase them on his own he may use a "hunting call" to attract other chimps. It's difficult for human observers to see, through all the leaves and branches, exactly everything that's going on during the chaos of a hunting raid in the treetops, but it appears that some chimps are blocking escape routes while others drive their prey to vulnerable locations where still other chimps are waiting to ambush them. After a successful kill, the monkey meat is shared among all the members of the hunting patrol, including those who were not directly involved in the final chase and capture.

Also in more controlled experimental settings, chimpanzees have demonstrated their capacity for cooperating with one another. Two chimps will each pull on a different rope simultaneously in order for both of them to get a banana or a sweet potato reward. Or one chimp will hand another a tool to gain access to a container with grapes.

We should not over-interpret the behavior of non-human animals, seeing in them what we humans would be thinking and doing. This has long been a contentious issue—whether what animals do is really the same sort of thing, with a similar kind of consciousness behind it, as what humans do, or whether humans are really unique in how they go about things. It may be that what to us looks like cooperation is really only individual animals looking out for themselves, even when they're in groups. Wolves may appear to be chasing a bison in relays, but it may be that the apparently intentional alteration of who's leading the chase might just be a change in the wolf that happens to be closest to the zigzagging bison at any given moment. Our impression that some chimpanzees are intentionally blocking a monkey's escape route for the benefit of the chimps who are doing the pursuing may just be the effect of the chance positions of the members of the hunting party. We might be reading into such patterns of behavior what we ourselves would be doing in that situation. Who knows what's really going on in the alien mind of a wolf or a chimpanzee?

When I cooperate with others, I think I know what's in my own mind. I know what to expect from the others in a cooperating group, and I know what they expect of me. I behave accordingly. The origin of this kind of mental connection is something of a mystery. Human cooperative relationships extend to people who are unrelated genetically, to groups larger than any biological family, even to complete strangers. This seems to be an innate trait, and a uniquely human one.

An infant just over a year old spontaneously hands back a pencil that an adult has dropped, or helps put clothespins in a basket, or plays ball with an adult, or helps clean up after lunch. At this age infants haven't yet learned how to talk, so their cooperative behavior would seem not to be learned, but innate, some basic part of being human. The urge to cooperate may be hardwired in the infant's brain as it develops in the womb, directed by genes inherited by all humans from their distant evolutionary ancestors a million years ago. The alternative is that in very young children cooperation and helping might nevertheless be the result of a socialization process. Parents and their infants do interact from very early ages, even without spoken language, the infant's developing cognition and motor skills prompting the parents to encourage and reward their cooperative behavior in nonverbal ways.

These two ideas are not mutually exclusive. There may be innate, gene-driven cooperation tendencies born in all of us, but such behavior is developed and refined only by social interactions. A study comparing families from Germany and India found that the frequency of helping acts on the part of the infant does in fact vary with parenting style, and maybe also with family size and the number of other children around. Therefore it looks as if socialization also plays a role in the development of cooperative behavior in humans.

Human-style cooperation—large scale, long term, including unknown participants, and involving a mutual understanding of the appropriate roles of everyone in the group—may have first arisen according to the usual evolutionary process of natural selection. Genes that directed cooperative behavior within social groups may have provided a selective advantage to group members; later, in wider yet more tightly knit human societies, such genes may have been aided in their transmission from one generation to the next by human teaching and learning capacities. The following explanation, while not proven, accounts reasonably well for the main facts of what we know so far about our ability to cooperate with one another in all kinds of useful endeavors.

The original genetic basis for cooperation among our early primate ancestors may have been "cooperation genes" that were shared by all the members of the family and that benefited them all. As populations increased in size, cooperative behavior could have spread to non-relatives because some family members made cooperative overtures to a few outsiders—a spillover from intra-familial cooperative behavior. Because of genetic variation among the outsiders, a few of them may have reciprocated. Those reciprocating outsiders naturally would have benefited from the cooperation. The original within-family cooperators would also have received extra benefits by cooperating with outsiders. Therefore, a genetic tendency to cooperate not only with family members but also with those beyond the bounds of family would have been reinforced. The non-cooperators from outside the family, tending to be shunned by the original familial cooperators, would have survived at lower rates and had fewer offspring. The original familial cooperators, in effect, would have been acting as selective agents for neighbors outside the family. Outside populations would then gradually have become more and more cooperative themselves. In this way "cooperative genes," like a beneficent virus, would have spread beyond family boundaries until they infected the whole population.

Meanwhile, our species continued to evolve stronger social interactions, which made cooperation easier, more effective, and more widespread. Cooperative behavior, which was originally supported only by genetically determined brain structure, began to be sustained also by early parent-offspring interactions, which led to more complex modes of cooperation. Genes and culture interact. In the same way, the language facility in humans, while innate, is in the earliest years rudimentary and unpolished, and becomes fully developed only with continuing interactions with adults.

In sum, the cooperative behavior observed in very young children before they have become "civilized" and seen also in our close evolutionary cousins the chimpanzees as well as in other primates, indicates that cooperative behavior does not come only from our culture and our ability to reason, but is built into our genes as part of our evolutionary heritage.

Altruism

According to the standard list in Christian theology, the seven deadly sins are pride, greed, lust, wrath, gluttony, envy, and sloth. (Judaism, Hinduism, Buddhism, and Islam have parallel lists.) Less abstractly, they can be translated, in order, as follows: "I'm the best, I want that, I want that, I hate you, I want that, I wish I had that, I don't feel like doing that." Even though they express natural tendencies toward self-preservation, that these self-centered feelings are labeled as "sins" could hardly be better evidence of the social nature of human beings. All religions urge us to be concerned with our fellow human beings over ourselves. We're all in it together. Ramana Maharshi, replying to the question, "How are we to treat others?" said, "There are no others."

One day my car broke down on a busy road. I steered it over to the verge, made the phone call, and waited for the emergency roadside service. When the truck arrived, the driver parked behind me, got out, and came over to see what assistance I needed. He stood at the very edge of the road, in some danger from the cars whizzing by. So I pulled him away from the road so we could talk more safely. What unconscious motivation was there for me to do this? The cynic might say that it was in order to avoid the inconvenience of having to call an ambulance and delay getting my car towed to a garage. That was certainly not in my conscious mind; I was concerned only for his safety, for his sake. Would this small gesture deserve to be called an altruistic act?

Betty also was expressing her helping predisposition with her *pro bono* legal work and her regular donations to the Save the Children Federation, Human Rights Watch, and the American Cancer Society.

Here are some real-life examples of other altruistic acts:

- In January of 1982, Air Florida Flight 90 out of Washington, D.C., its wings heavy with ice, failed to gain sufficient altitude after takeoff and crashed into the 14th Street Bridge, falling into the Potomac River. Two people on shore plunged into the icy water to try to rescue the few survivors. One passenger kept passing the helicopter's rescue line to other survivors in the water but drowned in his effort.
- In 1945, on the Pacific island of Iwo Jima, an American soldier threw himself on the hand grenade that had been thrown into his foxhole, shielding his fellow marines from the explosion. He died, but his comrades survived.
- A woman in a 400-meter hurdles race stopped in mid-race to go back and help a competitor who had hurt herself crashing into one of the hurdles.
- In the Kids Baking Championship, a time-limited cooking contest for 10- to 13-year-olds, a girl who was not quite finished stopped her own project to help another girl finish hers on time, thereby forfeiting her own chance to win.

This is all strange behavior, from the biological point of view of pure self-interest. Then there are the everyday acts of common courtesy: holding a door open for someone, offering your seat to someone, giving directions, letting someone break into line in traffic. Such minor gestures may have the same motivations and come from the same psychological sources as more dramatic altruistic acts, but they don't involve the same degree of risk or self-sacrifice usually used as the measure of "true" altruism.

Altruistic acts don't always have to be one-on-one as illustrated in Figure 56. They can also involve large, organized groups of people:

- In northern Japan, after the Fukushima nuclear power plant had been damaged by the massive 2011 Tōhoku earthquake and tsunami, workers and soldiers rushed in to protect people in the surrounding community from fires, explosions, and radiation leakage, exposing themselves to dangerously high levels of radiation in the containment effort.
- After the 2015 Nepal earthquake, aid organizations such as the Red Cross, Red Crescent, *Médecins Sans Frontières* (Doctors Without Borders), and many international teams of rescuers and volunteers joined in efforts to find and help survivors.
- In 2016, less than 24 hours after a mass shooting in a night club in Orlando, Florida, hundreds of people lined up to donate blood for the surviving victims.
- Cities throughout the world have shelters for the homeless, clothing drives, blood banks, organ donors, paramedics and ambulance services, police forces, firefighters, hospitals, and countless charitable and philanthropic societies.
- On the international scale, in addition to the Red Cross, Red Crescent, and Doctors Without Borders, there are also the World Health Organization, the United Nations Food and Agriculture Organization, the World Food Programme, CARE, Oxfam International, the Save the Children Federation, the Peace Corps, the International Rescue Committee, and many other international aid organizations—human altruistic activity on a grand scale.

The word *altruism*—placing the interests of another ahead of one's own—was coined by the 19th-century French philosopher Auguste Comte, from the Latin *alteri huic*—giving or paying attention "to this other one." One of the best examples of altruism in all of nature is the worker honeybee. She stings the intruding skunk, raccoon, bear, or human in order to protect the hive. But she loses her life in the act. How can the genes that underlie the making of her sting, the venom, and her behavior be transmitted to the next generation if she dies? And if in any case she is reproductively sterile? The answer is that those same genes are carried also by the hive's queen (the worker's mother), and the future queens of other hives (the worker's sisters), and the drones (the worker's brothers). So the genes are protected and passed on to the next generation by them, even if the self-sacrificing worker dies.

This genetic mechanism—natural selection of genes carried by the relatives who benefit—may be responsible, at least in part, for altruistic behavior in humans. We defend, care for, and have more loyalty toward members of our family and other close relatives. That can't be the whole story, because we behave altruistically toward strangers too. We give money to street-corner panhandlers. We donate to clothing drives, blood banks, shelters for the homeless, veterans aid organizations, and countless charitable and philanthropic societies. This wholesale, undiscriminating altruism is one of the more remarkable characteristics of human social behavior.

Is there in fact such a thing as true altruism, behavior that's truly based on caring about the welfare of another? Perhaps what appears on the surface to be an altruistic act is really, in the end, for our own benefit. Maybe we really care only about ourselves. It's possible that what we really want is to reduce the distress and the sympathetic anxiety we feel when we

Figure 56. Professional altruism, a noble calling (by Jules Cayron, 1868–1940; Smithsonian American Art Museum, gift of the Republic of France, 1915).

observe another's need or suffering. Or we want to avoid the disapproval of others, or we want to sidestep our own guilt feelings and self-censure when we *don't* help others. Or we unconsciously hope for some eventual reward by way of reciprocation, even if that reward is nothing more than relief from the obligation of helping.

These alternative explanations for why what looks superficially like an altruistic act is really self-serving have been tested in experiments in psychological laboratories. It turns out that even when people could easily evade a situation in which another is in distress, and avoid their own unpleasant sympathetic feelings, they nevertheless tend to stay and help the other person. Even when self-condemnation or social censure for *not* helping is removed from the equation, people still tend to want to help others in need. Finally, even when the possibility of a reward is minimized, people still tend to want to help others in the same measure. The alternative explanations do not hold up; we really are built to be genuinely altruistic. The adult human brain is indeed wired to generate helping behavior (at least under some circumstances) for the benefit of others, and not just for one's own benefit. We are not always total egoists.

The question remains: how does the human brain come to be wired this way? Is altruism something learned from parents and from society at large, so that early in childhood a degree of altruistic behavior becomes second nature? Or is it our *first* nature, something innate, part of our basic biology, behavior built into our genes and inherited from our evolutionary ancestors? The existence of altruism in very young children suggests the latter explanation. Children as young as 12 to 18 months of age, whose brains are still developing and who are hardly yet socialized, already show signs of being sympathetic to others' needs. They will spontaneously help an adult pick up something she has dropped or help her locate a lost article. Altruistic behavior patterns can of course be refined, channeled, and reinforced by early training and by example, but children are not just the self-centered little beasts upon whom parents, teachers, and religious leaders must layer a veneer of civilized behavior. Children, on occasion, can be innately helpful.

These two kinds of observation—very young children's behavior and psychological experiments on adults—suggest that altruistic behavior, even if not always evident, is part of basic human nature. This conclusion is nicely expressed in the Epistle of Paul the Apostle to the Romans: "when [they] which have not the law, do by nature the things contained in the law, these, having not the law, are a law unto themselves: which shew the work of the law written in their hearts...."

Mind-Reading, or Perspective-Taking

Human behavior can be described from the point of view of an observer floating high above the scene, looking down as if on a colony of ants, describing the ebb and flow of mass behavior without paying much attention to individual interactions. Yet such interactions do occur, and they are ultimately the basis for the collective behavior. These more personal, individualistic dealings, one on one, we can see only by getting inside individual minds. We humans can do that. We are good at imagining the beliefs, desires and intentions of other humans, their point of view, their sincerity or falsity, honesty or dishonesty, friendliness or hostility, trustworthiness or deceitfulness. In knowing also that others are good at this too, we are adept at deceit, at evading others' guesses about our own knowledge, desires and intentions—whether we're in a singles' bar, a suspect in an interrogation room, or a political candidate holding a press conference.

To bring to the fore what we all do unthinkingly, consider what you're doing when you glance in a wall mirror or a reflecting shop window as you pass by. You are checking yourself out: how do I look to others? What impression will I make? Would my appearance be attractive or acceptable to my interviewer, my boss, my client, the audience, a member of the opposite sex? You are momentarily taking another's point of view, another's perspective. A mirror is a perspective-taking device.

The ability to imagine what other people are or might be thinking or feeling—not necessarily always accurately—is variously called mind reading, mental state attribution, mentalization, social cognition, intention and behavior prediction, true and false belief understanding, theory of mind, interactional pragmatics, and perspective-taking. The last term best captures what's involved: taking the point of view of another person. Being able at least temporarily to take another's point of view is the basis for teaching, competitive sports, card-playing, and common courtesy. It lies at the heart of empathy and altruism. It is arguably the basis for effective communication and cooperation, and for human culture and civilization itself. Here are some examples of how it might be used—trivial examples, but they reveal day to day our automatic and unconscious perspective-taking:

- In the blink of an eye, something like this flashes through the mind of the tennis player as she prepares to serve: "My opponent knows that I know her backhand is weaker than her forehand, and therefore she might be expecting I'm going to serve the ball there; so if I serve instead to her forehand, I'm likely to get a weaker return than if my serve goes where she expects it."
- "I think you're having difficulty with your poems because you're looking at the words you've just written without hearing the sound of them inside your head. That's what I used to do. So try reading them out loud, and if you come to a place that doesn't sound right, stop and pay attention. Believe me, I know what I'm talking about."
- Said the patron to the waitress, "Too bad you don't have some extra arms." He momentarily felt her burden of having to carry so many plates at one time.
- Argentinian-born Paola says to another writer in the group, "There's a kind of contradiction in what you wrote. You're an experienced emergency room nurse and you've already been working here for a month and everyone knows you, so when the patient has just been wheeled in with a knife still sticking out of her, why would the staff only now say to you, 'Welcome to Miami!'?" The rest of the group immediately sees what the misunderstanding is: Paola thought it was a standard welcome, whereas it was an

ironic "welcome" that meant "now you see what you can expect in Miami." Every other person in the group, fleetingly taking Paola's perspective, can see what she was thinking, and understands why as a non-native English speaker she was missing the true meaning of the phrase.

In a remarkable study at the University of Plymouth in England, subjects were asked to quickly say whether an asymmetric image—such as the numeral "4" projected onto a table in different orientations from the normal upright image—was a mirror image or simply an image rotated in the plane of the tabletop: 4 ⇂ ⇃ ⇂ ⇃ ⇂ ⇃ ⇂. Response times varied from about half a second to a second or two. The subjects' responses were measurably faster when another person on another side of the table was viewing the same image in its normal orientation. It was as if the mental rotation that the subject had to carry out in evaluating the image was facilitated by the perspective of the other person—even when the other person was a purely passive participant, saying nothing. It's almost a kind of mind-meld, such highly social creatures are we.

Perspective-taking is an innate social skill that requires a degree of brain maturation before it appears. It develops spontaneously in children around the age of three years. The "Sally–Anne test" is one way of testing for it. In one version of the test, a child watches as Sally puts a marble in her basket. Then Sally leaves the room. Anne now comes over and takes the marble out of Sally's basket and puts it into her own basket. Sally returns, and the child is asked to indicate where Sally will look for her marble. Before the age of about three, the child will point to Anne's basket, knowing that's where the marble actually is. An older child, however, being aware that Sally *thinks* the marble is still in her own basket, will point to *Sally's* basket. The older child has developed a "theory" about the contents of Sally's mind, even though it's different from the child's own knowledge. Children younger than about three don't yet have this "theory of [another's] mind."

There are non-verbal versions of this test, which can be used to compare what chimpanzees and children can do in the way of taking another's perspective; i.e., of understanding another's belief even if the belief is known by the child or the chimpanzee to be false. In some experimental set-ups, chimpanzees repeatedly fail where children succeed. But in some circumstance chimpanzees (along with other great apes) do sometimes seem to have a dim awareness of the contents of minds other than their own. When, in view of the chimpanzee, a human actor hides an object and returns later to look for it again, the chimp will glance toward the place where it expects the human to look, even though in the meantime the chimp has seen the object moved to a different location. At some level, the chimp knows what's in the human's mind.

A telling incident at the Yerkes National Primate Research Center near Atlanta, Georgia, involving the chimpanzee Lolita and her new infant, was described by Frans de Waal:

> I was eager to see Lolita's baby, which had been born the day before. I called her out of the group and pointed at her belly [where the dark little blob that was her newborn was hanging on]. Lolita looked up at me, sat down, and took the infant's right hand in her right hand and its left hand in her left hand. This sounds simple, but given that the baby was clinging to her, she had to cross her arms to do so. The movement resembled that of people crossing their arms when grabbing a T-shirt by its hems in order to take it off. She then slowly lifted the baby into the air while turning it around on its axis, unfolding it in front of me. Suspended from its mother's hands, the baby now faced me instead of her. After the baby made a few grimaces and whimpers—infants hate losing touch with a warm belly—Lolita quickly tucked it back into her lap. With this elegant little motion Lolita demonstrated that she realized I would find the face of her newborn more interesting than its back.

Lolita knew the content of de Waal's mind, at least to the extent of knowing what de Waal would be interested in. Lolita was taking de Waal's perspective, reading his mind, and across species at that! (Quotation from *Our Inner Ape*, © 2005 by Frans de Waal, by permission.)

Taking the perspective of another individual, imagining what might be in another's mind, is an ability that is useful in a variety of social contexts, so it's not surprising that other social apes have that ability to some degree. Being able to picture what something might look like through other eyes may at bottom be an act of the imagination. Our superlative brain power allows us quickly, instinctively, and unconsciously to take note of subtle signs of voice, facial expression, gesture and body posture of the other person. Our brain is wired to mirror in ourselves the actions, intentions, desires, feelings, and moods of others, and so to participate in them.

The reaction of one human brain to what's going on in another's is the basis for our instinctive recognition of the authentic existence of others. It is the basis for teaching and learning by imitation, the interpretation of others' motivations and intentions. It is the basis of all sympathy, all altruism, and all morality and justice in human society. Each one of us participates in the same universal panoply of human sense impressions, the same basic human goals and desires. Because we are all swimming in our common biology, in the same fleshly pool, we all share in a kind of "distributed mind," akin to Carl Jung's "collective unconscious." Our biological and cultural cohesion makes us automatically feel and believe what others feel and believe.

It is a measure of the high degree of human sociality that we humans are better at perspective-taking than other social primates. More highly developed than in other primates, the human imagination is a mental facility that evolved especially in our branch of the primate evolutionary tree. It has enabled our species to construct advanced communication systems, advanced levels of cooperation, and collective inventiveness. That has made all the difference in our populating the Earth.

Evolutionary Origins of Sociality

The sections above suggest that specific elements of human social behavior—teaching and learning, cooperation, altruism, and perspective-taking—are instinctual and have a genetic basis. Is this true for human social arrangements in general, or do they have a more rational basis?

Human sociality could be the product of the human capacity for reasoning and learning from experience. In this picture, we humans rationally chose better social arrangements when we emerged from the primitive state of our ape-like ancestors and began living in larger groups. Writing a century before Darwin, Jean-Jacques Rousseau speculated that "morality began to appear in human actions, and every one, before the institution of law, was the only judge and avenger of the injuries done him, so that the goodness which was suitable in the pure state of nature was no longer proper in the new-born state of society" (*Discourse on the Origin and Basis of Inequality Among Men*, 1754); "only when the voice of duty takes the place of physical impulses and right of appetite, does man ... find that he is forced to act on different principles, and to consult his reason before listening to his inclinations" (*The Social Contract*, 1762).

On the other hand, in his *The Descent of Man* (1871), Charles Darwin was at pains to show that our moral sense—an important component of human sociality—is an instinct, inbred and innate. Other social animals also exhibit a concern to treat other members of the species well. And while in humans the individual's moral instinct is undoubtedly modified by the quality of parental upbringing and by social influences, there seems to be something about it that is not just learned but innate, as when we are emotionally moved by seeing a personal act of kindness toward a homeless person, or witness people cooperating to rescue a child who has fallen through the ice, or when we instinctually recoil from seeing a man mercilessly beating a horse. Darwin wrote,

> Selfish and contentious people will not cohere, and without coherence [cooperation] nothing can be effected. A tribe rich in the ... qualities [of fidelity, courage, sympathy, and trust] would spread and be victorious over other tribes; but

in the course of time it would, judging from all past history, be in its turn over-come by some other tribe still more highly endowed. Thus the social and moral qualities would tend slowly to advance and be diffused throughout the world.

Darwin recognized a problem in this simple account of social evolution. Those who helped others at their own expense might leave fewer offspring, and so genes for helping might gradually disappear from the population and the species.

> He who was ready to sacrifice his life ... rather than betray his comrades, would often leave no offspring to inherit his noble nature. The bravest men, who were always willing to come to the front in war, and who freely risked their lives for others, would on an average perish in larger numbers than other men. There-fore, it hardly seems probable, that the number of men gifted with such virtues, or that the standard of their excellence, could be increased through natural selection, that is, by the survival of the fittest; for we are not here speaking of one [whole] tribe being victorious over another.

He was nevertheless able to suggest a version that might work:

> Although the circumstances, leading to an increase in the number of those thus endowed within the same tribe, are too complex be clearly followed out, we can trace some of the probable steps. In the first place, as the reasoning powers and foresight of the members became improved, each man would soon learn that if he aided his fellow-men, he would commonly receive aid in return. From this low motive he might acquire the habit of aiding his fellows; and the habit of performing benevolent actions certainly strengthens the feeling of sympathy which gives the first impulse to benevolent actions. Habits, moreover, followed during many generations probably tend to be [come to be] inherited.

Darwin's proposal explains how social arrangements might have arisen as a genetically based instinct. Benevolent acts, if supported by behavior-modifying genes, and if reciprocated, redound to everyone's benefit. In modern terms, let's say that I, an uncommon member of a self-ish tribe, happen to have an uncommon gene that makes me sympathetic toward the plight of a neighbor, a cousin of mine. His roof has fallen in. I help him with the repairs. Because we're related, he carries that gene too, and later on he helps me shovel out the mud from my hut after the river overflows its banks. The long-term result might be that we both have an extra couple of kids who survive. Therefore, in the next generation copies of our genes will be present in higher numbers in the tribe than those of the average selfish member. Generations pass. "Sympathetic genes" increase in numbers and eventually come to characterize the whole tribe and eventu-ally the whole species.

Anthropological evidence tells us that sociality of one kind or another came long before there was such a thing as civilization. There never was a "pure state of nature" in which human beings acted only with their own selfish interests in mind. It's not true that only with civilization did morality in the form of concern for the welfare of others and a sense of fairness and justice come into being. Instead, human social arrangements—the precepts of our morality and religions, the organization of our corporations and governments, the structures of our legal and economic systems, our systems of education, our charitable organizations—all have an underlying evolutionary basis. They're all part of the biology of the social creatures we are.

On our own branch of the evolutionary tree, our ancestors were social creatures, although they may not have had all the social refinements that modern humans possess. Rousseau's vision of our past was wrong; human society did not arise full blown only with civilization. The evidence comes from three sources: first, modern anthropology, from which we can guess how humans may have lived several tens of thousands of years ago before the rise of agriculture and civilization; second, hints from fossils, which take us back a few hundred thousand years; and third, comparison of the species *Homo sapiens* with our other evolutionary relatives, which traces our social nature back a few million years.

The so-called hunter-gatherer cultures comprise over 300 small societies scattered over most of the continents of the world. They live without agriculture, without extensive private ownership of goods and property, and without the marked division of labor that has characterized modern society since 9,000 or 10,000 years ago. They typically live in small bands consisting of a few dozen related, close-knit, and emotionally bonded individuals in a small number of nuclear families. They are typically egalitarian, without definite long-term leaders and followers. Tasks are generally different for men and women: men fish, hunt for large animals, work stone, carve wood, make musical instruments, and conduct warfare; women care for children, make clothes, collect small game animals, water and fuel, and gather and prepare plant foods. But beyond these general tasks there is little division of labor. Strong cooperation among the members of these societies is the rule: women's care of babies and children is a community affair, food is hunted or gathered by small groups of individuals and is widely shared. Gift-giving, borrowing and lending are common, as is mobility and trade between groups. The opinions and customs of the group are important, and rules of behavior are strictly adhered to. Group discussions, storytelling, and singing and dancing are the warp and woof of the social fabric. There may be trading with nearby bands. If these modern hunter-gatherer cultures are anything like the way ancient people used to live, pre-civilization humanity was already highly social.

It's been said that we can't really know anything about the behavior and thought processes of ancient *Homo sapiens* from tens or hundreds of thousands of years ago, because there are no fossils of behavior or thoughts. That may be true, strictly speaking, but some archeological items can be *signs* of behavior and thoughts, Colored bead necklaces, small sculpted "Venus" figurines, symbolic markings on stones, and sophisticated cave and rock art from 25,000 to 75,000 years ago in Africa, Europe, Australia, and Indonesia all attest to the social nature of the peoples of those times and places. Such things are made by individuals not just for themselves. They are symbolic ways of communicating with fellow beings, just like today's jewelry and fashion and art shows.

Fossil bones from even earlier times also sometimes tell a story of the social nature of our evolutionary ancestors, even before our own species came into existence. A fossil found in a cave in Spain, dated to about 500 thousand years ago and probably belonging to *Homo heidelbergensis*, an ancestor of *Homo sapiens*, revealed a congenitally misshapen skull associated with facial deformity and probable mental retardation. Nevertheless, from signs of skull bone growth and the size of the skull, the child must have lived for 10 or 12 years. The child must have received a considerable degree of care before it died.

East of the Black Sea in the Republic of Georgia an even earlier evolutionary ancestor of ours, *Homo erectus*, left behind a well-preserved two-million-year-old adult skull and jawbone. All its teeth were missing except for one tooth in the left lower jaw. Extensive remodeling and repair of the bone around the tooth sockets indicated that the teeth must have been lost years before death. How did this individual survive? Gnawing the meat off scavenged bones would have been out of the question. If I were toothless, I would hope my friends would save me the ripest, softest fruit and soft animal brains, or at least that they would pre-chew my food for me. Maybe that's what his companions did.

These ancient ancestors of ours must have lived in social groups that gave their members extra attention and food when necessary, enabling injured or malformed individuals to survive. In many other species of ape, they typically would have been abandoned or killed.

Human sociality probably goes back even further. Our closest evolutionary relatives, gorillas and chimpanzees and bonobos, are all highly social. The vast majority of primates, from monkeys to baboons to apes, are social animals. We come from a long line of primate ancestors whose individuals interacted with each other and helped each other out in a variety of ways. With civilization—the domestication of plants and animals, the widespread sharing of agricultural practices and technology, the erecting of overarching sets of laws, and the establishment of culture-wide moral principles, religious beliefs and rituals, with language—we have of course added our own elaborations to our social practices, but it's built on a social primate foundation that goes back tens of millions of years.

The features that make human society different from that of other animals were added on late in our evolution. It is not a question of whether human sociality is behavior with deep evolutionary roots, or a recent invention. It is both.

16

The Truly Musical Animal

He stood beside a cottage lone,
And listened to a lute,
One summer's eve, when the breeze was gone,
And the nightingale was mute.
—Thomas Kibble Hervey, *The Devil's Progress* (1849)

In human evolution, music—defined broadly as rhythmic sounds with a distinctive pitch pattern—probably came before speech. Birds have it, whales have it, gibbons have it—although if these animals could speak, they probably wouldn't call it "music" in the human sense. In human languages—musique, música, musik, musiikki, mousiki, muzyka, musiqaa, tónlist, mele, sangeet, sagīta, ongaku, yīnyuè, eum-ag, umcolo, pūri—the word encompasses rhythmic patterned sounds that seem to have no utility other than giving pleasure.

In the dense scrubland west of San Francisco Bay a male song sparrow raised his voice in song—clear notes punctuated by sharp trills. He hoped to establish a territory and attract a mate. To that end he brought into play as many variations of his song as he could muster from the lessons he had learned from his father and his neighbors in his first year of life, adding a few changes of his own invention. Success! A female alighted on a nearby branch, cocked her head and listened, and then began, tentatively at first and then more explicitly, a display of her own. Wings all aflutter, she arched her back and raised her head and tail—her copulation-solicitation display, more public and shameless than would suit most humans. Over the next few weeks there followed egg laying, hatching, and hungry nestlings. So it has been for song sparrows for millions of years.

On a similar spring day on the west bank of the Garonne River in southern France, 800 years before that song sparrow was singing in California, a troubadour was singing a love song to any woman who would listen (Figure 57). His songs were based on what he had learned from his protector the viscount and from his friends, with variations of his own invention added.

> When the grass is new and the leaves emerge
> And the flower buds upon the branch
> And the skylark sings so high and clear,
> Raises his voice and sings his song.
> I feel his joy and the joy of the flower
> And the joy of myself, and more for my lady;
> From every quarter buoyed and bound by bliss,
> A joy that conquers every will.

Not recorded is how often his courting song prompted a human-style copulation-solicitation display from the object of his love (usually more subtle than that of the female song sparrow). From contemporary rumors, and the popularity of the art form, I presume that successes were frequent.

The word "song" might be used for any animal vocalization that sounds at least somewhat musical to human ears. From observation and experiment, the main functions of the songs of non-human animals fall into three categories: the attraction of mates, the defense of territory, and (in social species) the maintenance of cohesion in the group. Troubadour songs, the traditional playing of the guitar under the maiden's balcony, the military marches and songs that stir up a sense of patriotism and solidarity

Figure 57. Music, the language of love, a medieval troubadour (after "L'Improvvisante" by Francesco Maggiotto, late 1700s, from the Miriam and Ira D. Wallach Division of Art in the New York Public Library Digital Collections).

in the face of enemies, and the hymns and chants of religious gatherings, all illustrate the fact that the music produced by humans also serves these same three functions.

Animal vocalizations are limited in their biological uses and stereotyped in their performance. You wouldn't hear a whippoorwill wooing a prospective mate by opening his evening song with a new melodic line, or a meadowlark speeding up the rhythm of his song one morning to impress the others in the nearby field with his technical skill, or a goldfinch telling his nestmate, "You take the alto this time, honey, and I'll take the soprano." You wouldn't hear hyenas or seals or dolphins getting together to harmonize a new rendition of a popular tune for entertainment and the pure pleasure of it. Human song and human music have a wider range of uses than animal vocalizations.

Trained chimpanzees have been brought on stage to beat on a drum in accompaniment to human musicians. The chimps bang away, although not particularly rhythmically. The human musicians do their best to match the chimp's beat in order to make it sound something like intentional music. It's not the other way around: the chimp doesn't seem to have a real sense of rhythm or even the rudiments of what it takes to make human music. Only we humans have what we'd call real music: a true sense of rhythm, melody, and harmony, all of which we can vary at will. We make our music with our voices and with our pianos, drums, violins, trombones, and clarinets. The production of music by humans goes way beyond animal functionality; it comes from the human soul (whatever that may be) and is appreciated fully only by human minds.

We humans employ music everywhere in our social and cultural lives. Music is like a ring of keys that unlock different rooms of the soul: one for joy, one for gloom and foreboding, one for hope and relief from grief, one for religious awe, and so on. Music sets the mood for a story being told or an item being sold. Musical interludes plug up the empty intervals between radio and TV news reports. Music fills in the waiting times in phone holds. It's used just to get through the day or the occasion in a pleasant way, filling up otherwise unoccupied auditory space. There's elevator music, restaurant music, department store music, airport music, and the music of doctors' waiting rooms and dentists' chairs. Background music is everywhere in TV advertisements, adventure films, and nature documentaries. Some internet sites, TV channels, and radio stations are devoted entirely to music. Music accompanies most religious ceremonies; from gospel music to Fauré's *Requiem*, it can be religious in itself. It is a prominent part of funerals, weddings, coronations, and inaugurations. Music videos, Broadway shows, and operas tell their stories with music. The dramatic themes of motion pictures, whether love scenes or car chases, are almost always supported by music. Even in the era of the silent film live accompaniment was provided by a pipe organ or piano. Solitary individuals can be heard singing in the shower and whistling and humming to themselves in other locations. Even infants, before they learn to talk, dance to musical rhythms. We are truly an innately musical animal.

For humans, the making of music and the pleasure in listening to it seem far removed from the biological goals of surviving and having off-spring, the basic elements of evolution. Therefore, one can't help wondering where our music came from. A century and a half ago Charles Darwin wrote, "As neither the enjoyment nor the capacity of producing musical notes are faculties of the least use to man in reference to his daily habits of life, they must be ranked amongst the most mysterious with which he is endowed." Human music is more varied, more creative, more emotion-laden, more subtle in its rhythms, more playful, more purely pleasurable, and more enig-matic in its origins than the "songs" and "music" made by other animals. Watching the coordinated intricate fingering of the hands of a piano player, you can't help thinking, if you're of an evolutionary turn of mind, "*There's* a skill that could never have been used by our brutish ancestors for chasing down antelope prey, escaping from lions, picking fruit from trees, digging up roots, fighting off a competitor for a female, or repulsing the advances of an unwanted male. Where could it possibly have come from?"

Did our music arise gradually in some distant non-human ape-like ancestor from purely animal grunts and tool-making ability? Or was it a late development? Did our evolutionary ancestors become human first and only then start to make music? Or does our ability in the making and appreciation of music have nothing to do with evolution? Casting about for evidence of a whimsical god with some arcane reason of her own for mak-ing us the way we are, we could hardly do better than fasten upon her gift of music to humankind.

> "Music, the greatest good that mortals know,
> And all of heaven we have below."
> —Joseph Addison

When?

Religious musical traditions stretch back to the 9th and 10th centu-ries with the liturgical music and Gregorian chants of the Roman Catholic Church. Considerably earlier, in the Old Testament (Genesis 4:21, I Samuel 10:5, probably written around 600 BC), there are references to drum, harps, and flutes. Earlier still, the 3000-year-old writings, sculpture, and paint-ings of Egypt, India, and China describe harps, lyres, zithers, musical bells, flutes, and other instruments used to accompany dances, songs, prayers and religious chants.

Human music must be much older than written records tell, maybe even older than spoken language. There is a deep psychic connection between music and emotion, links perhaps forged during past evolutionary eons, built into the developing brains of infants as they are rocked in their mothers' arms and hear their mothers' lullabies.

The oldest undisputed remains of musical instruments, described in 2009 from caves in eastern Germany, were flutes fashioned from the wing bones of vultures and swans, dating back to about 40,000 years ago when human culture first began to flourish in Europe in the form of complex tools, representational and abstract art, small, sculptured figurines, ornamental beads and necklaces. Flute music might have been ceremonial, helping to maintain primitive groups' cultural identities. It's likely, even though we don't have the relevant archeological remains, that simple sounds were made with whistles and the rhythmic beating of drums long before more sophisticated music, possibly even before the evolution of *Homo sapiens* 300,000 years ago. Whistles and drums might even have been made and used by an ancestral member of the *Homo* genus, the adept tool-maker *Homo ergaster*, two million years ago.

How? The Muses Weigh In

Early humans may have first made music by using their voices. Instrumental music may have been invented simultaneously, or perhaps even earlier with the discovery that useful and pleasant sounds could be made by banging on things, or blowing into them, or by plucking a string of a hunting bow.

The word *music* comes from the Greek μουσική, *mousike*, originally meaning *any art or science presided over by the Muses*. They were traditionally nine in number. Later usage restricted the word to the art of what we call "music" alone.

There are several possibilities for the origin of music, which is essentially a continuous string of sounds organized into patterns of rhythm, pitch, and intensity. It might have first come into being as an extension of prior human behaviors, which can be classified according to the predilections of five of the Greek Muses (Figure 58). Music might be an extension of dance, of language, of more primitive forms of music itself, of natural sounds from nature, or an inevitable historical development of human culture. A separate case can be made for each of these apparently contradictory ideas, as outlined below.

Terpsichore — *dance*   Calliope — *epic poetry*   Euterpe — *music*   Thalia — *idyllic poetry*   Clio — *history*

Figure 58. The origin of music is disputed (images modified from "The Nine Muses" by A. Rey and P. Bineteau, circa 1844–1861, Miriam and Ira D. Wallach Division of Art, the New York Public Library).

1. **Terpsichore, the Muse of Dance.** First there might have been dance, rhythmic movements of the body. Then voice in the form of chanting was added, and eventually instruments. Even though some late-developing branches of dance are less rhythmic—modern interpretive dancing, for example—dance was almost certainly rhythmic in the beginning. The developing human brain seems to be designed by evolution to be receptive, even before birth, to the rhythm of the mother's heartbeat. After birth, the newborn's innate sense of rhythm is reinforced by the mother's rocking the baby to sleep.

Rhythm for humans is not just a regular beat but one that can be varied at will, faster or slower, with hierarchies of stronger and weaker beats—such as, in one of its simpler manifestations, three-quarters waltz time. Other animals do make sounds with a regular beat, but they don't have the ability to vary their rhythms and make different combinations of them, changing their vocalizations or other sounds from simple rhythms to compound rhythms. You've heard a woodpecker drumming on a tree, but you've never heard drumming like this: ta,ta,ta,ta…taTA,taTA,taTA,taTA,taTA,taTA… TAtata,TAtata,TAtata,TAtata…and so on.

Even before *Homo sapiens* appeared, our pre-human ancestors may have developed a special sense of rhythm when they began to walk and run with a regular cadence. Our natural rhythm, the usual "andante" tempo, arose from that. So rhythm was already built into our anatomy and our nervous system from the start. Human babies have a sense of rhythm. Very young children have a sense of rhythm and dance even before they can talk.

It is primarily rhythm that gives us pleasure. The rhythmic beating of a drum without change in pitch can be enjoyable, making us feel as if we want to get up and dance. A string of sounds that consist merely of changes in pitch, without rhythm, as might for example be produced by a clarinet, is likely to be unpleasant.

We are able to dance together in a coordinated fashion when there's a regular rhythm to follow. This gives us a sense of other people, a sense of the group, a sense of belonging—one of the earliest traits that defines what it is to be human. In war dances, hunting dances, and other ritual and ceremonial dances throughout history and throughout the world, you can see and hear the primitive origins of music. In the marching exercises in military basic training camps, you can witness the ancient dance rhythms.

What about other animals? There is the regular drumbeat of hooves in horse races, the regular wing beats of small birds, the courtship dance rituals of the bird-of-paradise, the grebe and the flamingo, the rhythmic cooing of doves and portions of the songs of goldfinches. Rhythmic all these may be, but these animals produce it automatically and don't seem to have the sense of it. They don't vary it, they don't develop it further. Only we humans use rhythm as the basis for developing our complex, intricate, and varied musical compositions—worthy of the gods, one might say.

2. **Calliope, the Muse of Epic Poetry, which uses language to tell stories**. First there might have been language. It is language, after all, that sets humans apart, as does music in the way that humans make it. Once language had reached a certain level of complexity, a natural rhythm began to emerge. (Just read the last sentence out loud.) The human mind could have fastened on that, and the first poet was born. Changes in pitch were added to the rhythm, and songs arose. In this way the structure of music came to be modeled on the existing patterns of language. Where language has syllables, words, and sentences, music has notes, phrases, and themes. A limited number of syllable sounds can be put together in different combinations to create tens of thousands of words and an almost infinite number of sentences. A limited number of notes, determined by the hearing range of humans and standard ways of dividing the musical scale, can be combined to make an endless number of melodies. As dialects and new languages evolve, so do new musical styles and cultures. It was language that first moved humans to ecstasy or tears; the emotions provoked by music are echoes of the power of language.

3. **Euterpe, the Muse of Music**. There might have been music before there was either language or dance. There was singing—music without words—consisting of primitive vocalizations that proclaimed territory, attracted a mate, maintained group cohesion. Hearing musical sounds conveyed to the hearer the mood and intention of the vocalizer. While these

vocalizations developed refinements and elaborations to become what we call real music, only later did they branch out to become ordinary language in its more precise, more communicative, and informational mode.

Mothers talk to their babies using the words in the language the mother knows. But on all continents, whatever the language, this "motherese" is spoken more slowly, and the mother's voice is more highly pitched, more articulated, and more singsong-y than the speech adults use to talk to one another. It is not the words that the infants attend to; it is the pitch contour, the rhythm, and the emotional quality of the mother's voice—her song. It is not the information content of the mother's cooing that counts, but her caring and her comforting presence that her voice conveys. The human infant appreciates the musical elements of the mother's voice long before it achieves any language ability. Only later does it grow out of infancy and begin to understand the meaning of words. In fetal development, the music appreciation part of the human brain develops first; it is a more ancient part of the human brain than the part devoted to speech.

Besides, many animals make melodic sounds intended to convey feelings of one sort or another, and obviously they do so without language. So music originates from a source older than language.

4. Thalia, the Muse of Idyllic Poetry, which draws its inspiration from Nature and the pastoral life, including the sounds that animals make. First there might have been only animal sounds. These were later modified along two separate tracks in humans: one leading to articulate speech that communicated information, the other leading to music with its initially ritualistic, group cohesion, and mate-attracting functions, and later as a form of art and pleasure. It may be that neither language nor song, however defined, was the original *source* of music. The original universal "sounds of music" were the sounds of animals, back when the human ancestors were animals. Gibbons sang in the forests. Dolphins, whales, and seals sang under water. Birds sang from the fields and the treetops. Even words may have begun as human imitation of animal and bird calls—a useful skill in ancient hunting practices, and one still used by hunters in Africa, the Americas, and Australia. Human-style music with its varied rhythms, melodies, and harmonies might have been an invention based on these sounds. Whatever similarity there is between music and language, it is because both are the daughters of animal sounds.

Human music is more like animal sounds than it is like human speech. Like many animal vocalizations, human songs repeat single notes; they combine small numbers of notes into short phrases, which are also repeated, with variations; and they have a hierarchical structure in which larger motifs and themes combine notes, phrases, and their variations.

Pitch and note intervals are important in both animal sounds and human music, but not so much in most human languages. Human language with its words and syntax belongs to humans alone, while it is human music that connects humans with their animal past.

Many of the neuronal pathways used in music-making and music appreciation are different from those used in spoken language. Brain damage from a stroke that impairs language ability may nevertheless leave victims retaining their musical abilities. And the converse: some stroke victims who have lost their sense of melody are still able to converse normally. There is also a genetic condition (called "congenital amusia" or tune-deafness) in which the ability to distinguish pitch and melody is defective, but language facility is normal. These observations suggest while musicality and language share some of the same brain pathways, they evolved separately. Musicality is a distinct human trait.

5. **Clio, the Muse of human History and Culture**. First there might have been human culture. Many tens of thousands of years after the origin of our species *Homo sapiens*, long after communicative speech was in place, art for its own sake arose—carved figurines, cave paintings, rock art. Music in its present form was an outgrowth of human sociality and human creativity that developed along with other art forms. Music combines natural sounds, language, ritualized chants, and dance. Its origin had nothing to do with biological evolution but arose when humans developed an aesthetic sense for the first time and came to appreciate beauty of all kinds: beauty in a sunset, a waterfall, a hummingbird, a cave painting, as well as in fashioning pleasant sound-producing instruments, and adding the human voice to it. Music has no biological purpose other than to give humankind a sublime kind of pleasure. As Joseph Addison wrote, "Music, the greatest good that mortals know / And all of heaven we have below."

Human music is *made up*. It's made up from the simple notes first found here and there in nature and then copied. But it's more than just a collection of those simple notes. It's a new creation with new properties and new effects. Music is made up in the same way that mathematics is made up from simple numbers. It's made it up in the same way that from the basic elements of human experience fairy tales are made up, which we then use to terrify and charm ourselves.

Human musicality is therefore a creation of human culture, which could only have arisen when humans had reached a cognitive level where their intellectual, emotional, social and spiritual life became as important as physical life.

The Muses have had their say. There are no strong reasons for preferring one view over another, except for a qualification of Clio's contention that music arose only with the development of human culture. Given that there are special regions and nerve pathways in the human brain devoted specifically to music-making and music-appreciation—brain functions that are not present in other primates—it's reasonable to conclude that there are indeed evolutionary, not just cultural, underpinnings of human musicality. The cultural vs. evolutionary hypotheses can be reconciled by saying that human musicality originated as an evolutionary process and was then refined by cultural advance. Otherwise, some of the abilities evolving in conjunction with dance, language, and imitation and elaboration of animal vocalizations could all have been involved in the progression of our musicality. As for the reasons for its evolution, music remains, in Darwin's words, "amongst the most mysterious" of our faculties—although there's nothing wrong with a little creative speculation.

As Part of Our Biology, Why Music?

The "why?" question is an evolutionary question. Applied to music, it can be translated as follows: *What evolutionary advantages—that is, better survival and reproduction—are gained by having musical ability?* There might be different kinds of evolutionary advantage, depending on the kind of music, such as military march music or romantic crooning. In addition, whatever the advantages that might have prevailed in the early history of our species, and that might have been responsible for the early establishment and spread of musicality, they are not necessarily the same ones that keep music going in our own times.

The mystery of music is part of the evolutionary mystery of art in general. What is it *for*? By what evolutionary process did it become established as such an integral and universal part of human life? Watching and listening to Ana Vidovic play guitar in concert, you have to think that what she's doing has nothing to do with her biological function as a woman. Now divorced from its origins and appreciated for its aesthetic and mood-altering qualities, music in the beginning (contrary to Clio's assertion) may have had specific biological functions, echoes of which might sometimes still be dimly heard through the din of modern culture. There are four areas in which early musicality in our evolutionary ancestors might have contributed, directly or indirectly, to success in survival and reproduction: attraction between the sexes, the establishment and defense of territory, group solidarity, and the raising of offspring. Clues about the origins of human music come from observing what other animals do.

Here are Charles Darwin's thoughts on the matter of attraction between the sexes:

As the males of several quadrumanous animals [that is, "four-handed" apes and monkeys, with both the hands and feet having opposable thumbs] have their vocal organs much more developed than in the females, and as a gibbon, one of the anthropomorphous apes, pours forth a whole octave of musical notes and may be said to sing, it appears probable that the progenitors of man, either the males or females or both sexes, before acquiring the power of expressing their mutual love in articulate language, endeavoured to charm each other with musical note and rhythm.... The impassioned orator, bard, or musician, when with his varied tones and cadences he excites the strongest emotions in his hearers, little suspects that he uses the same means by which his half-human ancestors long ago aroused each other's ardent passions, during their courtship and rivalry.

In stark evolutionary terms, the hypothesis is that those ancestors who made good music and succeeded in charming the opposite sex had more offspring. Single examples don't prove the point, but they can illustrate it: Fats Domino had 11 children, Ray Charles had 12 children (by 10 different women), B.B. King is reported to have had 15 to 18 children (by several women), and Johann Sebastian Bach had 20 children, 10 of whom survived to adulthood. By contrast, accomplished female singers such as Maria Callas, Birgit Nilsson, Leontyne Price, and Celia Cruz had no children, while Ella Fitzgerald, Billie Holliday, Nina Simone, and Barbara Streisand had only one each. The discrepancy between the sexes demonstrates the biological principal that sperm cells are more numerous than egg cells; they are cheap and can be spent freely without hindering a musical career, while fertilized egg cells and all they entail are expensive and must be limited if the female is to have an illustrious career.

Besides enticing the mate you desire, the other side of the male courtship coin is warning off potential rivals. If you're a songbird, your superior musical voice is not only attracting your female partner but is also saying, "This is my territory. Keep out." That birds use their song in this way has been shown experimentally by capturing a male bird (such as a song sparrow) and removing him from his territory, after which he is replaced with a speaker that rebroadcasts a recording of his song. As a control, some territories are left unguarded. Empty territories are taken over by other male song sparrows, but when a territory has a speaker playing the song, it is not invaded by any other song sparrows. (It could be imagined that a self-confident singer like Enrico Caruso might be sending a similar message to would-be rivals for female attention, but playback experiments like those with the song sparrows haven't been done.)

Group vocalizations might also have had an early function in territorial defense, an important social behavior in animals that protect their resources for foraging, as chimpanzees do, or for hunting, as wolves and lions do. Group calls and cries would signal to potential invaders, first of all, that it is in fact a *group* that the invader has to contend with. A synchronized chorus could moreover indicate to potential invaders that it is a *coordinated* group, one acting as a unit, and therefore more formidable. Individuals genetically predisposed to participating in such choruses would have a better chance of passing on their genes than an individual who is roaring or howling on its own. It's conceivable that war chants of early humans might have had this function.

Another early advantage of having music was likely the sense of community it brought when performed or listened to with others. Even before the appearance of *Homo sapiens*, this practice might have had evolutionary value in survival and reproduction because of its use in supporting cooperation in hunting, food sharing, and defense. You can expect help not only in return for the help you actually provide others in the group but also in the demonstration of your willingness to provide it. Any behavior that ties you to the group's activities in which you willingly participate for its enjoyment will ultimately redound to your benefit. Your participation gives you, in the long run, a better chance of surviving and of having more offspring. In this way, music in any one of its various forms functions in the same way as religion. Whoever has been part of a choir, chorus, orchestra, folk dance ensemble, garage band, marching band, or ballet troupe, however large or small, hasn't felt connected and uplifted by being one of the performers?

Related to the sense-of-the-tribe role of music and dance are work songs, the songs of farm laborers, rowing-boat songs, and sea shanties. These are highly rhythmic, sung in unison to relieve the boredom of repetitious work ("Tote that barge! / Lift that bale! …"). Before the industrial age and even today in some parts of the world, work songs make the tedious work of rowing, chopping, pounding, hoeing, hauling and stowing more tolerable and in fact more efficient.

> The J.M. **White** is a bran' new **boat**
> Stem to **stern** she's mighty **fine**
> Beat any **boat** on the N'Orleans **line**
> Stowing-sugar-in-the-**hold** be**low**
>   *Hey, ho, be***low***, be***low**
>   *Stowing-sugar-in-the* **hold** *be***low**
>   *Hey, ho, be***low***, be***low**
>   *Stowing-sugar-in-the-***hold** *be***low**

Songs like this, besides advancing the work, benefit each individual in the group. In the long run, each individual reaps the rewards of the group effort.

Music and rhythm also have positive if subtle effects on newborn babies and young infants. Mothers use lullabies and rhythmic rocking to calm their infants. Even in its first year of life, well before it develops language capacity, an infant is already sensitive to the rhythm, pitch, tempo, and melody of its mother's speech and songs. The calming effects on the infant can be observed even when it's a recorded song, and even when it's a mechanical cradle that's doing the rocking. So the infant brain must already be tuned to the appreciation of some of the essential features of music. It's significant that this is the case. Newborn babies are sensitive to being touched, and to tone of voice and facial expressions of the mother—but why *music*? Music seems to act as a milk supplement with its own measurable and positive effects on the infant's growth and well-being. Perhaps in some profound way lullabies reproduce and reinforce the caring behavior and comforting murmurings of the mother.

In our ancient evolutionary ancestors, genes predisposing mothers to sing to their newborn babies would have been passed on to future generations in greater frequency if those babies had a better chance of thriving than babies who were not sung to. In mechanical evolutionary terms, newborn human babies are like empty boxes outfitted in advance with small drawers and cubbyholes ready to receive the nourishing musical notes that the singing mothers pour into them. Those boxes thrive better than boxes with featureless interiors and silent mothers; they grow up to be effective adult music-boxes in their own right. Genes for singing and genes for receiving song are preferentially propagated together down the generations.

Music can also provide benefits for adults. Music affects mood and can soothe the troubled mind. It can assuage feelings of loneliness. It can produce greater tolerance for pain. The field of music therapy relies on the restorative powers of music performance and listening, and the positive emotions they engender, providing a supportive background for healing and health. Music can reduce stress, anxiety, and grief. Music can elevate the spirit. A neutral state of mind can be lightened or darkened by music.

All these positive effects of music for the propagation of what could be called "music appreciation genes" might help account for the present musicality of *Homo sapiens*, this most cooperatively social of species.

Given the world's diversity of musical styles and genres and the variety of uses to which music is put, perhaps the best view, the one that embraces all these ideas and observations, is that music is not one thing with one origin, but many things with many origins. Dance and rhythm may have started early, with origins in the body's natural motions. Animal vocalizations with pitch changes, with its multiple functions in establishing territory, attracting mates, and promoting group cohesion, may have come even earlier, gradually morphing into song, retaining some of their original communicative uses and adding others as spoken language and the expression of emotions wove themselves into human messages. With tool-making and the advance of technological skills, instrumental music came into its own, having earlier shadowy origins in rhythmic drumming. Some kinds of music and dancing retain connections with these ancient origins. Others are elaborations that have severed their relations with their original evolutionary advantages, just as poetry is now divorced from imitative hunting sounds, and as abstract art is from early cave paintings. The different forms of music may have originated for different reasons and at different times in humans' long evolutionary history, but today we lump them all together into the catch-all category of *music*. No wonder we haven't yet successfully pulled out any single thread from all these entanglements.

In spite of its pervasiveness in human culture and human psychology, in the overall scheme of evolution music may have played, and may still be playing, a relatively minor role in human survival and reproduction. It does have measurable effects in helping new mothers adjust to their new lives with their new offspring and in calming their babies, and in soothing troubled minds, altering mood in positive ways, and reducing the thresholds of pain. But music is not as essential for life as is digestion or wound healing or the beating heart. Music does not construct the various biochemical pathways on which life moves forward. But it smooths them out a little and that may be sufficient reason for its existence and pervasiveness in human life.

Whatever genetic basis there may be for human musicality, culture certainly plays a major role in determining individual musical tastes. In Appendix II you can find a list of a few of my personal favorites.

17

The Talkative Ape

...words that flutter and agitate thought...
—Virginia Woolf, "Fishing" (1937),
in *The Moment, and Other Essays*

The framers of us framed the mouth, as now arranged, having teeth and tongue and lips, with a view to the necessity and the good contriving the way in for necessary purposes, the way out for best purposes; for that is necessary which enters in and gives food to the body; but the river of speech, which flows out of a man and ministers to the intelligence, is the fairest and noblest of all streams.
—Plato, *Dialogues* (*Timaeus*; 4th century BCE)

Relaxing in my favorite coffee shop, I was idly listening to a group at a nearby table who were engaged in a ceaseless, relentless chattering, like the twittering of nervous birds. The language was foreign, so the meaning of what they were saying didn't interfere with my wonder at the raw phenomenon.

What was going on? At a guess, they were describing what happened earlier in the day, things they saw, projects they hoped to complete in the future. They were sharing their joys and their anxieties. They were giving opinions, making requests, making plans, dispensing or asking for advice, gossiping, conveying information and ideas, agreeing and disagreeing.

Our speech not only mirrors our minds, but it is also the warp and woof that holds our social fabric together. The ancient emergence of language must have gone hand in hand with increasingly sophisticated social interactions and with the rise of complex social institutions. It is the means by which we persuade and deceive others around us for our individual benefit and by which we induce their trust and cooperation for our mutual benefit. It is also the means by which our experience is vicariously widened beyond the limits of our narrow individual horizons.

Human language is so complex, so subtle, so varied, so unlike anything possessed by dumb animals that Johann Peter Süssmilch in 1766 could write that it must have been a divine gift to humankind. There is such harmony between the intention of the speaker and the response of the hearer that this double gift could only have come from the gods. Further, since the human ability to reason depends on language, and language depends on reason, neither of which is much use by itself, the gods must have given us this other dual gift as well. It couldn't have been derived from a mere elaboration of the instinctive animal sounds.

Yet other human traits such as the eye, the kidney, and the shoulder joint are also complex structures, and we view them as structures that came into their present state of being by gradual evolutionary processes. Why could not the complex structure of the larynx and tongue, and all the brain circuits that control them, have arisen in the same way? Several special problems present themselves to the evolutionary biologist. There are no obvious fossil traces of language that would allow the reconstruction of an evolutionary history. There are no examples of proto-languages among living animals, no simpler transitional forms of language that suggest the gradual stages by which human language might have reached its present state. On an evolutionary timescale, human language may have appeared rather suddenly, not gradually. And what is its reproductive value? Given how well mice and rabbits reproduce themselves without language, can we say that having language helped the survival and reproduction of ancestral humans in any way? Do lines such as "How do I love thee? Let me count the ways. / I love thee to the depth and breadth and height / My soul can reach…" result in more children being born?

For a more thorough discussion with many details and examples of language and its origin, the reader may profitably read Steven Pinker's masterful book *The Language Instinct*. Nevertheless, brief and oversimplified as is the treatment here, the reader may find it useful as an introduction to the subject.

The Anatomy of Speech and Language

Language is the major mental instrument we employ for apprehending reality and communicating our conception of it. What is required for vocalizing the words we want to use is, first, the airway and the mechanical adjustments of the sounds issuing from it: the lungs and diaphragm, the windpipe (trachea), the voice box (larynx) with its cartilage supports, muscles, vocal ligaments (vocal cords), the hyoid bone to which the ligaments are attached, the tongue, lips, teeth, and the various muscles used in coordinating their use and in breath control. Second, there are the brain circuits—centered, in most people, on the left side of the brain—necessary for controlling the airflow from the lungs and all the laryngeal, tongue and facial muscles involved in speech production. There are, as well, the parts of the brain that set the whole apparatus in motion when you have something to say, some emotion to express, some command to give, some comforting words to convey. As in bowling, golf, or tennis, there are the mechanics of stance, body and arm motion, and swing; and there is the brain's decision about where to aim the ball. Language is the brain's tool; speech is the act of using it.

From observations of the effects of strokes and other localized brain injuries on language ability, along with studies using magnetic resonance imaging (MRI) and other physical methods on conscious subjects, we have learned that some regions in the left cerebral cortex are especially important for language ability. These regions are located in and around the Sylvian fissure, the groove that separates the temporal lobe and adjacent regions of the frontal and parietal lobes (see Figures 49 and 59). They include the classic Broca's and Wernicke's areas—named after Pierre Paul Broca and Karl Wernicke, the French and German 19th-century physicians who separately described the language impairments that result from injuries and disease in these two brain regions. The observations mean that there are brain regions specifically devoted to language production and language comprehension, separate from basic intelligence, which is not much affected by these conditions. Damage to Broca's area in the brain usually results in a deficiency in speech *production* without any loss of understanding of what others are saying or what she wants to say ("expressive aphasia"); an example might be the following speech of a woman trying to explain why she went to the emergency room: "Yes … ah…. Monday … um … hospital … and…. I can't … um … nine o'clock … and oh…. Thursday … ten o'clock … um … yes … doctors … two … and doctors … and um … car…." Damage to Wernicke's area, on the other hand, causes problems in language *understanding* ("receptive aphasia"). Patients have difficulty both in understanding others and in producing sentences that make sense, while their speech itself may be grammatical, rapid and fluent, but without understandable content (for example, "You know we steam with the water kind of play happy at the moment and that we had cats and dogs to get him round cats and dogs at the moment and the point is to take care of him like you want I want before…"). Broca's and Wernike's areas in the brain are close to, and have interconnections with, other regions that receive hearing input from the inner ear, and regions with motor nerve output that controls movements of the muscles of the tongue, mouth, and larynx.

Both the speech anatomy and the controlling brain regions are

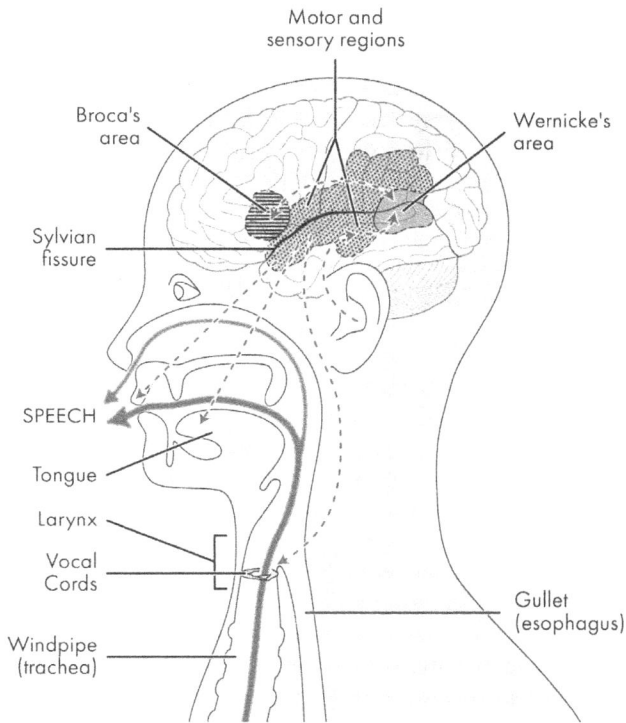

Figure 59. Language pathways are near the Sylvian fissure on (in most people) the left side of the brain. Broca's area is important for effective speech production and Wernicke's area for language comprehension. Other important brain areas are sensory regions that receive hearing input, and motor regions that control speech, such as the muscles in the lips, tongue, vocal cords, and other muscles. Dashed arrows represent schematically (not anatomically) nerve pathways.

necessary for language; they must have evolved together to produce the human language ability.

Features of Human Language

What sets human language apart from the communication systems of other animals and makes humans unique among all the species of the Earth? It is its variety and its fine-grained structure with all the different meanings that can be attached to the same sounds arranged in so many different ways, corresponding to the fluctuating focus of the human mind and its diverse perceptions, preoccupations, and desires. Human language has four basic features not found elsewhere as a package: it is combinatorial, symbolic, syntactical, and hierarchical to degrees not found in the animal world. The closest noises we might find in other animals are the calls and songs of birds, the squeaks and squeals of dolphins and whales, and the various vocalizations of chimpanzees, which contain some of the elements found in human language but none of them by themselves developed to the degree that they are in humans.

The language of humans is primarily a system of vocalizations that communicate information, feelings, intentions, exhortations, and imaginings. Sometimes it is mere wordplay. We are able to say rapidly, *Peter Piper picked a peck of pickled peppers* and *Seven slick slimy snakes slowly sliding southward*—an ability that requires the rapid-fire machinery of the whole vocal tract, the activity of the vocal cords, breath control, movements of the tongue and lips, and all the nerve circuits in the brain dedicated to determining the content and meaning of a given act of speech and initiating and coordinating the necessary movements of the different parts of the anatomy to produce it.

(Apologies to professional linguists for the overly simplistic and imprecise definitions that follow.)

- *Combinatorial.* Language combines sounds in different ways to give different meanings: *step, pets*—same sounds, different order, different meanings. For another example, the seven phonemes ɑ ɛ ə e i ɪ r (symbols of the International Phonetic Alphabet for six vowel sounds and one consonant) can be combined in different ways to produce different words with different meanings: *are, err, era, ear, air, airy, aria, area, eerie, eyrie, rare, ray, rhea, rear.* Humans can easily make 50 or more different sounds with their vocal apparatus (the total number of vowels and consonants varies somewhat with the language) and combine them to produce tens of thousands of different words.
- *Symbolic.* There are many onomatopoetic words such as *whoosh, crack, boom,* and *chickadee*; some have speculated that the earliest words of our ancient ancestors were all of this kind, words that sounded like what they referred to. But most words are symbolic only: their sounds have no relation to the things themselves. There is nothing at all "tree-like" in the words for *tree* (English), *arbol* (Spanish), *derevo* (Russian), *shù* (Chinese), and *rakau* (Maori). The word *malevolent* is not evil sounding at all, but rather smooth and pleasant to the ear.
- *Syntactical* (Greek, *syn*, together, and *taxis*, arrangement). The order of words matters; different arrangements have different

meanings. The same four words are used, but *Jim killed the alligator* doesn't mean the same thing as *The alligator killed Jim.* Rules also exist about how the words in a sequence must be related to one another in order to convey meaning accurately; these rules are called "grammar"—which is related to syntax but encompasses additional rules. *The apple doesn't fall far from the tree* is grammatical, whereas *The apple fall doesn't from far the tree* has the main words still in the same order, but the relations of the words don't follow the rules, and the sentence is ungrammatical.

- *Hierarchical.* Language is built up from a limited number of small sound units (phonemes) to make syllables and words, then phrases, then whole sentences. The initial sounds are meaningless; in combinations they make meaningful words; when the words are combined with other words to make phrases, the meaning changes; and when the phrases are combined, the meaning changes again.
  - ◉ *sounds* (roughly corresponding to the written letters): a c d e f g i m n o r s t v ea ou ng th ti
  - ◉ *words from combinations of these sounds*: air, are, at, coming, conditioning, eerie, from, of, rear, sounds, stage, the, there, vents
  - ◉ *phrases from these words*: at the rear, coming from, eerie sounds, of the stage, the air conditioning vents, there are
  - ◉ *a sentence from these phrases*: There are eerie sounds coming from the air conditioning vents at the rear of the stage.

Another feature related to language's hierarchical structure is the embedding of a sentence or clause within another sentence. In *She believes her sister is a thief*, "her sister is a thief" is embedded in the principal sentence "She believes...." There can be several levels of embedding: *I think she believes her sister is a thief.* And *Mr. Baldwin was at a loss to explain why Annette had decided to tell him that his sister was having an affair with a man Annette thought was his brother*—embedding with a vengeance. It strains the brain a bit, but with a little mental effort the sentence is nevertheless coherent and understandable. It requires the hearer's or reader's mind to temporarily store several ideas and then quickly retrieve them in the correct order as the sentence structure dictates, keeping the bookkeeping straight so as to understand the relations being described. (Similar mental processes are also required to understand, at first encounter, single words such as *disingenuousness* and *noninterference.*) This mental ability is unquestionably beyond the mental capabilities of non-human animals, but it's a common feature of human language.

Some animals—birds, hyenas, dolphins, monkeys, chimpanzees—use different vocal signals that mean different things to them. These vocalizations don't rise to a level even approaching human language. But they show that some of the basic elements of language can arise in any moderately sophisticated social animal. Our language facility, of course, could not have been inherited from these animals; birds, hyenas, dolphins, monkeys, and chimpanzees are not our ancestors. Nevertheless, our not-yet-human evolutionary ancestors no doubt possessed some of those language features too.

Precursors of Language in Animal Communications?

Extracting meaning out of vocalizations is such a useful ability that it should not be surprising that it has arisen in some form or other in other animals—especially among animals that live in social groups. Meaningful vocalizations are produced by birds, hyenas, dolphins, monkeys, and other apes besides the specialized ape *Homo sapiens*.

Many birds send messages to other birds by means of vocalizations that we're pleased to call "songs." These consist mostly of one or a few stereotyped melodies that indicate either *I'm here and ready for mating* or *This is my territory, stay out*. Considerably more versatile are the different calls of crows, which vary in pitch, duration of the caw, intensity, number of repetitions, and duration of the intervals between repeats. The different kinds of *caws* appear to mean (i) *look out for danger from above*, (ii) *there's danger below*, (iii) *flee!* (iv) *calling all family members*, (v) *it's me, not some stranger*, (vi) *come on, there's food here*. Other calls may have other meanings, depending on the situation. Such calls could be regarded as symbolic, each conveying a particular meaning.

"Mobbing" and harassing a predator is behavior common to many birds. When my cat goes outside, the local mockingbirds gather and swoop down and scream at it with a distinctive cry. Without even seeing my cat, I know when it's prowling around outside. A single bird may mob my cat, but its mobbing call may attract other birds who will join in. The highly social pied babbler of South Africa is a medium-sized bird with a white body and head, and a black bill and black wings and tail. It uses, among several other calls and notes, a mobbing call in the presence of a snake, fox, or hawk. The mobbing call consists of a longish squawk or squeal followed by series of short, rapidly repeating harsh notes, in a pattern like this: /---/-/-/-/-/-/-/. Interestingly, the two parts of the mobbing call, /---/ and /-/-/-/-/-/-/, are also used separately. The first part is a "soft alert" indicating either that a harmless animal such as a rabbit is nearby, or that a predator is somewhere in the neighborhood but is not an immediate threat; the second part is a "recruitment call" indicating that nearby members of the group should gather together for foraging. Combine the two and the bird has a mobbing call: alarm plus gathering. In experiments in which various calls are recorded and played back to the birds, only this one elicits the mobbing response. So these two notes together, in this order, have a specific meaning for these birds—an example of a vocalization that uses a simple kind of syntax.

The African grey parrot is remarkable (for a bird) for its ability to imitate short phrases of human speech and for its ability to use them in the appropriate social context: *Hello!* when its owner enters the room; *Goodbye, love you*, when its owner leaves the house, and *Where are you?*, when its owner is in the house but out of sight. Whether this ability is true speech of a simple kind, or merely imitation prompted by the social circumstance and dependent on an excellent memory, is not clear. In the wilds of its native western African jungles, this highly social bird is not going to be found conversing with its neighbors in English. But it may employ its imitative abilities and vocal agility to identify itself to its fellow parrots as an individual or as a relative, and to convey its location, where to find fruits and seeds and a good roosting place.

Some of these vocalizations are probably just species-specific noise, to which other birds (and other animals) may respond. Other vocalizations carry more information about the individual and the circumstance and could be considered symbolic in that they have specific meanings that young birds carry within them as an instinct or learn from their parents.

Among mammals, spotted hyenas have different kinds of whoops, grunts, giggles, growls, whines, laughs, and squeals to communicate different messages depending on the circumstances. The vocalizations are threats or warnings, general alarm calls, indications of social rank, casual greetings, greetings after a long absence, announcements that a prey animal is in the vicinity, and calls for cooperation in a hunt. These are useful communications, but they are not true language in the human sense, because the different sounds are not assembled in different combinations and arrangements to make more complex meanings.

Dolphins use whistles, squeaks and clicks to communicate with one another. They have "signature whistles" by which they identify themselves as individuals to other dolphins nearby. They can echo-locate objects, distinguish one object from another, and report their findings to fellow dolphins and to their human trainers. The dolphin Akeakamai at the Kewalo Basin Marine Mammal Laboratory in Hawaii learned to follow specific gestural commands and to associate a trainer's gestures with objects and locations in the tank. She learned to interpret the different meanings when the same gestures were delivered in a different sequence. For example, having learned the tank's gesture-language—that modifiers precede the object and that the gesture for the object to be moved precedes the action command— Akeakamai would watch hand signals in the order *surfboard-person-fetch*, and would promptly bring a person (in the water) to the floating surfboard; but with the hand signals given in the order *person-surfboard-fetch*, Akeakamai would bring the surfboard to the person. She could react correctly to a location signal incorporated into an object-plus-command sequence: in response to the gesture sequence *left-ball-hoop-fetch*, she would transport a hoop to a ball on the left side of the tank. With two balls and two hoops in the tank and given the gesture sequence *right-hoop-left-ball-fetch*, she would unhesitatingly and correctly transport the ball on the left side of the tank to the hoop on the right side. How exactly dolphins in the wild might use this mental ability, this apparent appreciation of simple syntax, is not known.

Vervet monkeys give different alarm calls according to whether the danger is from a hawk, a snake, or a leopard. That there is real meaning in the different calls is demonstrated by the way the hearer responds, dropping to the ground if it's a hawk alarm, or climbing a tree and hiding in the foliage if it's a snake or leopard alarm. Something like such vocalizations, present in our ancestors, could have been the first "words," although such word-object associations are a long way from combinatorial, hierarchical, recursive language.

Chimpanzees, our closest living evolutionary relatives, living in their natural habitat in the forests of central Africa, use mixtures of gestures, vocalizations, and facial expressions in their communications with one another. They grimace, compress their lips, stamp their feet, wave their arms, gesture with their hands. They vocalize with pant-hoots, whimpers, screeches and other calls to convey what might mean in English *Stop that!* or *Give me that!* or *Come closer,* or *Don't bite me!* or *Let's have a grooming session,* or *Follow me,* or *Let's go.* The chimpanzee communicative repertoire is moderately wide, and subtle to a degree, although it doesn't rise to the level of human language.

Washoe, a female chimpanzee who died in 2007 at the age of 42; Kanzi, a bonobo; Koko, a gorilla; and Chantek, an orangutan, were all able to communicate with humans by using gestures of American Sign Language (ASL), and even better by learning to press symbol-bearing keys on a keyboard. Chimpanzees can link ASL gestures and keyboard symbols to actions and objects in the external world—a mental ability that is an element of human language. This ability comprises just a simple vocabulary, not a complete language. It lacks the versatility, content, precision, subtlety and grace of human language. And different arrangements of gestures and symbols don't mean much to a chimpanzee or a bonobo, although Kanzi could respond correctly to the commands "put the ball on the blanket" and "put the blanket on the ball." The appreciation of the distinction might be a recognition of simple syntax, but it could also be simply a matter of reaching first for the object named first.

At the Yerkes National Primate Research Center, a visitor asked one of the researchers, "How is it that these chimps can use ASL at all? It's not part of their nature, and they never use it in the wild, do they?" She replied, "No, but how is it that humans can play the piano? That's not part of our evolutionary heritage either. Just like with humans, chimpanzees' natural abilities can be refined and guided by training."

The conclusion from these observations on animal communications is that there are some simple elements of language that are widely shared across many animal groups—use of the voice to make sounds that serve as labels for predators or for particular social situations, and in a few species, minimal kind of syntax in which a different combination of two vocalizations has its own meaning. Human beings, as animals, share these basic abilities too. But no other animal goes on to combine dozens of different sounds in different ways to make tens of thousands of different words, which themselves can be arranged in different ways to make an almost infinite number of sentences. No other animal has tens of thousands of different combinations of sounds and words, elaborate syntactical arrangements, and hierarchical and recursive constructions. No other animal has language.

Weaver birds show some construction skills when they use grass and twigs to make their nests, and orangutans break leafy branches and bend them to make their sleeping platforms. But only humans have refined their abilities to a point where they use steel, aluminum, concrete, and glass to build skyscrapers. With the way we arrange the sounds we make with our voice, we can express subtle shades of feeling, share our ideas concrete and abstract, bind ourselves together (or separate ourselves) with words and sentences that create trust or suspicion, dominance or submissiveness, that convey sincerity or craftiness, loyalty or treachery, love or hate.

An Evolutionary Pathway for Language

We don't know when our evolutionary ancestors became capable of spoken language. Certainly, there was no definite time before which they didn't have language, and after which they suddenly did. There was no lightning strike, no clap of thunder. Human language could not have appeared all of a sudden with all its present complexity. For one thing, to judge by other anatomical changes that have occurred over the course of human evolution, thousands of years would have been required for the evolution of all the necessary neuromuscular and pharyngeal anatomy that is peculiar to humans. Also required would have been the evolution of modified brain circuits for the production and comprehension of speech. This probably accompanied an increase in general cognitive ability—an understanding of cause and effect, an ability to rearrange objects in our environment, and words in our brain, in order to accomplish a given outcome or solve a particular problem. (As mammalian brain function goes, understanding the preceding sentence is no mean feat.) The brain and the larynx must have learned to cooperate in arranging different sounds in different orders to make different words and sentences. From sounds heard, the brain must have created meanings. There followed hierarchical structuring and syntax, phrase embedding, all the while accompanied by an increase in our ancestors' mental ability to analyze and interpret what was going on in the world around them.

Human speech and language must have developed gradually, and such an elaborate structure could not have arisen by accident. Unless the gods somehow intervened, natural selection must have been operating in some way. The underlying gene changes that must have occurred in our evolutionary ancestors were of course random and accidental, but if any of those mutations chanced to have had positive effects, however slight, on survival and reproduction. Their representation in the population is likely to have increased as the generations went by, eventually resulting in human language as we know it now. What were those positive effects? How could evolution have occurred in such a way as to change an ape capable of making only a relatively small number of meaningful sounds (perhaps something like those of modern chimpanzees) into an ape with a full-blown combinatorial, syntactical, hierarchical, grammatical language?

There are no truly primitive languages surviving today, ones that might give clues about what early stages of language development might have looked like. All modern languages—from those used in Moscow or London or Beijing to those spoken in the jungles of the Amazon, on the Aleutian Islands, or in the Kalahari Desert—are complicated; they are all combinatorial, syntactical, hierarchical, and grammatical.

Evidence of language in our early evolutionary ancestors is hard to come by. The earliest definitive written records go back only about 7000 years. They were just marks on clay tablets, intended to keep accounts. If accounts had to be kept, prior transactions must have occurred. Complex and meaningful sentences must have been spoken a long time before anything was written down, perhaps thousands or even tens of thousands of years earlier.

Archaeologists haven't found any fossil tape recorders nor heard voices still faintly echoing off the walls of ancient caves. But a few paleontological and archeological observations suggest that some semblance of human language might—just might—have existed 500,000 years ago. One of our likely ancestors from that time, *Homo heidelbergensis*, had middle-ear bones and the hyoid bone (to which muscles of the tongue are attached) very similar to those of modern *Homo sapiens*, suggesting that they had the hearing and vocal capabilities of modern humans. Unfortunately, that observation doesn't count for much, because careful analysis of the vocal tracts of monkeys shows that even they could have language and produce passable speech, if only they had the brains for it. What we modern humans seem to have over other primates is precise brain control of the vocal tract, and the ability to learn how from our elders (we do learn, with great dexterity, whatever language culture we grow up in).

More telling are other archeological discoveries. An essential element of language consists in having a mental concept of symbols—understanding that something, even a sound, can stand for something else. Symbolic elements are commonly found in prehistoric art and other objects; they could be a sign that the humans who produced them had the mental capacity for language. The oldest cave- and rock art in Europe, Africa, and Australia dates back to about 30,000 or 40,000 years ago. The art portrayed not just the local lions, bears, rhinos, and elk, and the outlines of hands, but also more abstract compositions of lines, dots, and whorls. These may have been merely decorative, but they also may have been a form of communication between the mind of the artist and that of the viewer.

Archeologists have uncovered evidence that around the same time in Africa and Europe people began using red ochre, an earthen clay rich in iron oxides, as a pigment for decorating pottery, beads, weapons, and even the bones of the deceased. They might also have used the pigment for painting the body for ritualistic observances. Such practices might have had symbolic meanings in encouraging success and good fortune or warding off evil.

Even older, from 70,000 years ago, are the pierced marine shells and pieces of the shells of ostrich eggs found in Blombos Cave in South Africa. They almost certainly served as beads for necklaces. One can imagine a young woman wearing such a necklace, which would have announced symbolically to the men of her village, "I am the chieftain's daughter; you'd better be someone special if you want to mate with me!"

It's possible that the cave art and the beads and the body pigments of ancient humans were mere decorations, indications of a developing aesthetic sense. On the other hand, they may be signs of an awakening symbolic thinking that accompanied the development of real language.

Anatomically modern humans, *Homo sapiens*, first appeared in Africa about 300,000 years ago. For the next 200,000 years or more, according to the kinds of stone tools they left behind, there was little change in behavior. Then during the period between 70,000 and 40,000 years ago, things began to change, apparently earlier in Africa than in Europe. In what has been called "The Great Leap Forward," or the "Rise of Behavioral Modernity," more advanced stone tools appeared, along with bone tools, signs of trading between groups, increased cooperation and greater division of labor within groups, more sophisticated art and decoration, and music. This was likely the time when more complex human language began to emerge.

According to the evidence from fossil skulls there was no change in overall brain size during the rise of behavioral modernity. The question is whether the increase in technological and societal innovation, including language, could be accounted for by non-genetic changes such as increased interactions between people as populations spread from one geographic region to another and as they developed more trading relations, or whether there were underlying, gene-based brain changes that could have led to the development of the language facility. It could have been both.

A few tens of thousands of years is a short time when compared with the several million years of evolution of the genus *Homo*, but it is long enough for new gene mutations affecting brain function to arise and become established by natural selection. Already present from our earlier primate ancestry would have been brain associations between particular sounds and particular objects or events, as well as a minimal capacity to learn by imitation. New genetically-based brain changes could have allowed a variety of sound combinations—that is, words and word combinations or syntax—to refer to a greater range of objects and events in the world. A gradually increasing cognitive ability, useful in understanding the world generally, could have eased the understanding of the meanings of the new sound combinations. As long as these abilities could be transmitted to the next generation, by ordinary gene inheritance and by parents teaching their offspring, the increase in mutual comprehension and sharing of useful information would have benefited everyone who could participate, giving better survival and more children. Their wordless and grammarless forbears would have been left behind. Language facility (although not any particular language) would have come to be inherited and would have evolved toward increasing mutual comprehension and information-sharing. In this way the human facility for language could have evolved, providing the basis for further social and cultural advances starting about 40,000 years ago.

Now our words and sentences stream into us from others, and we ourselves pour them out into the air for others to hear and react to. They are invisible, evanescent, and we imagine them to be like gossamer threads that exist for the moment and then vanish. They are in fact the steel beams and girders around which societies and civilizations are built.

18

Human Races,
Real and Imagined

> The question whether mankind consists of one or several
> species has of late years been much agitated by anthro-
> pologists, who are divided into two schools of mono-
> genists and polygenists. Those who do not admit the
> principle of evolution, must look at species either as sep-
> arate creations or as in some manner distinct entities....
> —Charles Darwin, *The Descent of Man* (1871)

Tramping around in a steamy mangrove swamp on the north coast of
Jamaica in my heavy rubber boots and long-sleeved shirt, doused in mosquito repellant, I looked up to see three local boys watching me from several yards away. They were about 10 years old, in shorts, and barefoot in the
mud. I hadn't realized there was a small village nearby. "Watcha doin', mis-
ter?" said one, without a trace of self-consciousness or mistrust. "Just looking for insects," I said somewhat guardedly. "Can we help?" he asked. "Show
us what kind."

In the 1980s I was making a survey of the Caribbean islands in search
of new races of a species of termite that lived in dead mangrove trees. I
found them, with the help of the boys in the swamp and the laboratory
facilities of my new colleagues at the University of the West Indies—most
of whom were black, as that category would be defined in the United
States. It wasn't long before the nebulous sense of black-white racial tension, which in Miami you could feel floating quietly under the surface, had
noticeably dissipated. Here in Jamaica the community I belonged to was
less concerned with skin color but had common interests in doing science.

The Jamaican insect populations had some traits which were slightly
different from the same species on the U.S. mainland. In most traits it's
recognizable as the same species, but under a low-power microscope you
could see that the protuberances on the front part of the head were a little

225

more prominent, the third antennal segment was slightly longer, and there were more bristles on the back of the thorax. A higher-power microscope revealed that some chromosome segments were arranged differently in the Jamaican insects.

None of this was especially unusual. Any species spread over a sufficiently wide geographical area is bound to show racial distinctions. In northeastern United States, milk snakes have an average of 198 scales on the underside, and those in the southeast average only 175 scales. Yellow wagtails in Russia have gray feathers on the top of the head; in Egypt those feathers are black. The northern Siberian tiger has a thicker and paler coat than that of the slightly smaller Bengal tiger in the south. The chimpanzees of central Africa differ slightly in their anatomy, behavior, and DNA sequences from the chimpanzees in western Africa 1500 miles away; the central chimps are on average slightly larger, their arms are slightly longer, their brow ridges a little less prominent, their faces and bodies slightly less hairy, and the skin on their faces and bodies is darker.

The Concept of Race

Are human races a biological reality or just a social construction? The answer is *yes* and *yes*. The idea of *race* runs on two tracks, the biological and the sociological. Among animals and plants and humans, and in any species distributed over a wide enough geographical area, distinct geographic populations, or races, are a biological reality. In humans their boundaries have been blurred by wars, conquests, and migrations. As a consequence of the evolutionary history of our species over the last 300,000 years, the boundaries were in fact indistinct to begin with. The boundaries between races were indistinct but nevertheless still real, still discernible in patterns of gene combinations from different parts of the world. On top of that reality, human nature for its own social purposes has erected racial categories that have only tenuous connections with biology.

Just as racial differences are a fact of life among insects, snakes, wagtails, tigers and chimpanzees, so they are a fact of human life. This is only to be expected from our species' having spread widely across the surface of the globe during the last hundred thousand years or so. Human beings could not have escaped the evolutionary processes that produce geographic differences. The history of our species unearthed in archeological digs, in the relationships of different languages, and in DNA sequences, demonstrates that the different populations of today did not have independent origins (see Figures 61 and 62). They have the same kind of partitioning of differences that we see in other species. The racial differences are apparent in the human population of Jamaica, reflecting the history of the island. Eighty to 90 percent of Jamaicans have African ancestors (mostly slaves brought in by the British in the 18th century to work the sugar plantations) or are descendants of unions between Africans and Europeans; the remainder are mainly of European and Asian descent.

If there really are such entities as human races, what are the biological relations among them? Or is the concept of human races a mere social construction without a significant biological basis?

There are social constructions that are regarded as "races," as the first three columns in the table below show.

U.S. Census	Brazil Census	British Police	Some Mid–20th Century Anthropologists
Asian	Amarelo (Yellow)	Asian	Mongolian
White	Branco (White)	White	Caucasian
Black	Preto (Black)	African	Negroid
American Indian	Pardo (Brown)	Indian/Pakistani	Amerindian
Pacific Islander	Indigeno (Native)	Mediterranean	Australian
Hispanic/Latino			

The categories in the first three columns in the table were constructed primarily for social purposes. Because of their history of slavery and conquest over the last few centuries, many Latin American countries informally add other racial categories to their self-descriptions: mestizo, mulatto, moreno, caboclo, zambo, Quechua, Aymara, Guaraní, European, Afro-Bolivian, Afro-Colombian, etc. Such classification schemes have only a tenuous connection with biology. The fourth column, a five-race classification, was an attempt (now outdated) to bring some sort of rational, biological order to the differences between the peoples of the world. (Other anthropologists, attempting a finer-grained analysis, preferred six racial categories, or seven, or even nine, adding Polynesian, Melanesian, Micronesian-Papuan, and East Indian to the other categories.)

The word "Caucasian" refers to the Caucasus Mountains, a mountain range between the Black Sea and the Caspian Sea on the southern border of Russia, in a region that is now occupied by the countries of Georgia, Armenia, and Azerbaijan. Johann Friedrich Blumenbach, an early and influential German anatomist and physiologist, wrote in 1795 in *De generis humani varietate nativa* ("On the Natural Variety of Mankind") as follows:

> I have allotted the first place to the Caucasian, for the reasons given below, which make me esteem it the primeval one ... [this variety of mankind has] that kind of appearance which, according to our opinion of symmetry, we consider most handsome and becoming ... that stock displays ... the most beautiful form of the skull, from which as from a mean and primeval type, the others

diverge by most easy gradation on both sides to the two ultimate extremes (that is, on the one side the Mongolian, on the other the Ethiopian). Besides, it is white in colour....

Blumenbach also quotes the 17th-century travel writer Jean Chardin: "The blood of Georgia is the best of the East, and perhaps in the world; I have not observed a single ugly face in that country, in either sex; but I have seen angelic ones. Nature has there lavished upon women beauties which are not to be seen elsewhere."

A classification of five races of supposedly independent origin is illustrated in Figure 60. The old 20th-century anthropological view was, in the main, that the major human races arose from some shadowy past, populating Europe, Africa, Asia, the Americas, and Australia independently. Their physical and biochemical differences involved adaptations to conditions on the different continents: dark skin presumably to provide protection from the burning African sun; light skin the better to absorb the weak sunlight of northern Europe and counteract the effects of diets relatively deficient in vitamin D; an extra skin fold to shield the eyes from the cold dust-laden gales of the Mongolian steppes; a narrow nose in the dry, chilly north to moisten and warm the air before it reaches the lungs; blood proteins to defend against parasites common in Africa. Beyond these, what other innate differences might lie buried in this racial categorization was anyone's guess. Perhaps there were fundamental differences not just in physical traits but also in intelligence, inventiveness, energy and drive, honesty, morality, passivity or assertiveness, gentleness or propensity to violence and savagery—suppositions for which the evidence is lacking. A frequent assumption underlying many writings and social policies was that light-skinned people are the best.

The Evolution of Human Races

Most 20th-century anthropologists were working with inadequate information, without the results of DNA surveys and without a broad evolutionary viewpoint. They were constructing categories that did not reflect the true status of human variation any better than did the social categories. The old anthropological race categories were wrong on two counts. They didn't show the true genetic relations of the different peoples of the world, and they embodied the erroneous idea that human races have well-defined boundaries, each with its own set of traits beyond what could be seen with the naked eye, such as a rebellious nature, acceptance of rigid traditions, a love of drinking or dancing, belief in magic—traits that were often assumed to have a genetic basis but which were in fact cultural.

Figure 60. An outdated concept of five main human races formed separately on different continents.

The DNA sequences of human populations contain an evolutionary history. In the same way that evolutionary trees can be constructed from differences in physical traits and genes (see Figures 5 and 21), so evolutionary trees can be made from the variability in the DNA sequences of different *races*. This within-species variability comes from random DNA mutations that inevitably arise over time and as populations of a species spread into new geographic regions.

A new population derived from an older one brings its ancestral group's DNA sequences along with it and adds a few new random DNA variants of its own. (Without such DNA differences, tracing one's ancestry to different parts of the world would not be possible.) Most new DNA changes have no effect on how the body functions except for a few DNA changes related to adaptations to new environments. With the aid of computer programs to help sort out all the branching events, the data on DNA sequence variation in the world's human populations, along with archeological findings, give the history of the whole human species. This is shown in Figure 61.

The branching pattern of humanity's spread throughout the world can be straightened up and re-represented without the world map behind it. The result is shown in Figure 62, which illustrates the genetic relationship of the human populations of the world.

The old-fashioned racial classifications do not apply. They have been discarded because they are not relevant to human evolutionary history as we now know it. Figure 62 shows that Africans are not a single group,

HUMAN MIGRATION & POPULATION EXPANSION

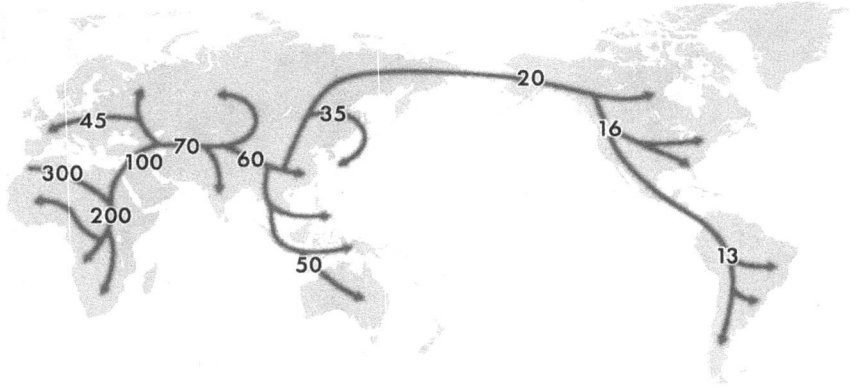

Figure 61. The spread of *Homo sapiens* from its origin in Africa to other parts of the world, a map based on genetic analysis and archeological discoveries. The numbers indicate the approximate time (in thousands of years before the present) when our species first reached that position on the world map.

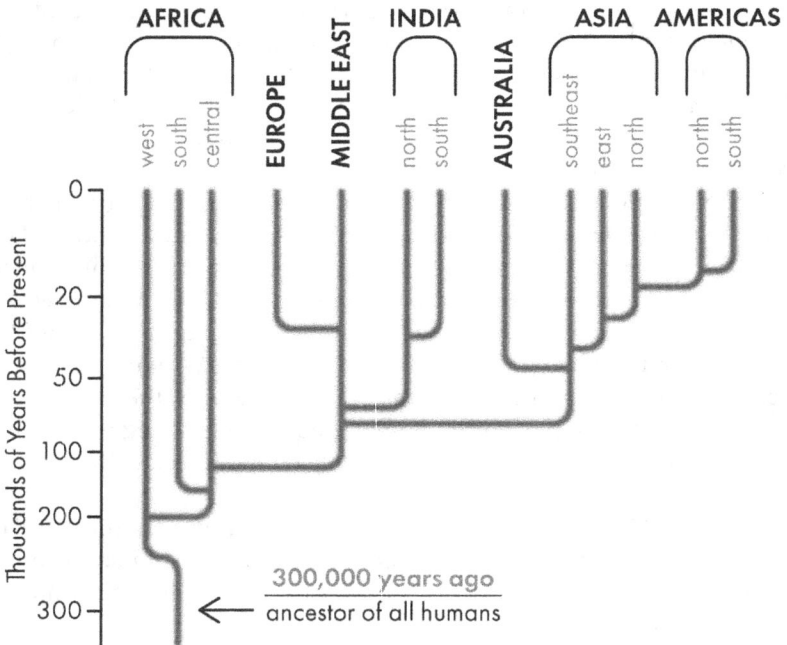

Figure 62. The relationship of different human populations, based primarily on genetic analysis.

but comprise several distinguishable populations, more genetically distinct from each other than are people in the rest of the world. Europeans and Middle Easterners are two branches of people who were once a single population. Populations that lived in China branched out into Australia, eastern and northern Asia, and the Americas; they therefore make up a group of genetically related populations that could be called a race of "Easterners" if one liked, to be distinguished from all the people further west from which they originally came. Native Americans do not have to be considered a separate race of people; they are an offshoot of northern Asian populations.

Some populations had past connections with neighboring populations (not shown in the diagram). DNA from two adjacent populations mixed together when one population migrated into a region already occupied by the other and they had offspring together. Even discounting the conquests of the last two or three thousand years and modern world travel, many of today's geographic populations, because of ancient migrations from one region to another, are not completely homogenous. Europeans, for example, are not a pure population; they contain mixtures of DNA from several-thousands-of-years-old waves of migrations from Africa, from the Middle East, and from northwestern Asia. In the United States the terms "European," "Caucasian," and "White" are used interchangeably. From a population-genetics point of view they have different meanings: the first refers to people of mixed ancestry who have occupied Europe for five or six thousand years; the second refers to people from the region of the Caucasian Mountains separating Europe and Asia; and the third refers to self-identified light-skinned non–Hispanic people in the United States who claim ancestry mostly from western Europe. The difference in the points of view provides some justification for the claim that the idea of "race," socially considered, is just an artificial construction.

Many Africans themselves carry small amounts of DNA from Middle East farmers who migrated back into Africa three or four thousand years ago. Even with these migrations, population differences persist because people did not freely interbreed across wide geographic regions, but tended to keep to themselves, choosing mating partners near where they lived. The different populations are like viscous puddles of differently colored paints, blending into each other at the boundaries, but still retaining traces of their origins at their centers.

What have been called *populations* above can just as well be called *races*—groups distinguishable not only by a few physical features and DNA sequences, but also by their ancestral relationships. These don't provide an unambiguous guide as to what defines a *race*. The word *race* is nevertheless useful in limited biological contexts, just as the words *game* and *religion* are useful even though they cannot be precisely defined. They all refer to something real even if fuzzy at the boundaries.

From the branching patterns in Figure 62, can we say how many human races there are? Perhaps two: a major branch, Africans, and another major branch, everyone else, the non–Africans. Or maybe three: Africans, Middle Easterners plus Europeans, and everyone else who does not belong to those two. Or maybe seven, or even 13 if we want to regard every branch in the diagram as a definite, distinguishable group of people on genetic grounds. Or even more, if we take into account further subdivisions that are not shown in Figure 62, but which would emerge from more detailed genetic analysis. Racial distinctions can be made on the basis of DNA sequences reflecting the history of the human species, but the boundaries are arbitrary.

But on genetic grounds there are no sharp boundaries between groups, just as there are no sharp boundaries between the different colors of a rainbow, even though we find that different names for the colors are useful. An illustration of the fuzzy genetic boundary between two races (or populations)—Middle Easterners and western Asians—is given in Figure 63.

In a rainbow, pure colors of distinct wavelengths do exist, but in *Homo sapiens* there is no such thing as a pure race. Traces of different geographic populations from the earliest days of our species a hundred thousand or more years ago are still distinguishable on the basis of their DNA sequences. But with all the forward- and back migrations and all the intermixing and interbreeding that have been going over the last few tens of thousands of years, all the old populations are noticeably adulterated. In the human species there is no definable "pure race" anywhere in the world.

As for intelligence, as measured by scores on various kinds of IQ test, there is no good evidence for genetic differences between races. For individuals *within* a race (however defined), IQ differences turn out to be partly the result of gene differences between people (and partly the result of their differing environments during development), but that is not the same thing as claiming that average differences between whole populations are the result of systematic genetic differences between those populations. Some of the many variable genes that influence performance on IQ tests do vary in frequency between populations, but such differences tend to average out when the effects of all the different genes are added up. The result is that the overall gene influences on IQ between populations are negligible. The evidence suggests that the differences in the small but measurable differences in average IQ scores between populations are not based on gene differences between them but on differences in social, economic, and cultural conditions.

How much genetic difference overall is there between the different human populations, compared with the genetic differences between us

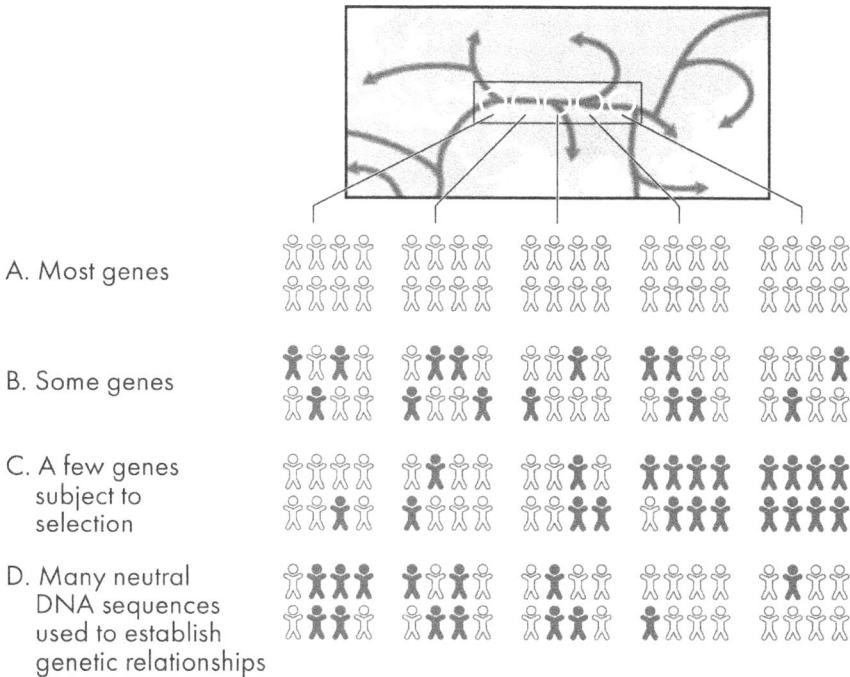

A. Most genes

B. Some genes

C. A few genes
   subject to
   selection

D. Many neutral
   DNA sequences
   used to establish
   genetic relationships

Figure 63. A schematic depiction of the indistinct boundary between two human races, Middle Easterners represented by the white circles on the left side of the map and Western Asians represented by the white circles on the right side. Eight individuals from each of five regions from west to east across the boundary summarize the results of genetic surveys. Open figures are people carrying ancestral genes/DNA sequences; filled figures are people carrying new (variant) genes/DNA sequences. Genes and DNA sequences of people from these five regions fall into four categories: A, top row: most genes are the same in all populations; B, second row: some genes are variable throughout all populations; C, third row: some genes, such as skin-color genes, have undergone natural selection (adaptation) as human populations spread from west to east; D, fourth row: many neutral DNA sequences are variable from population to population in such a way as to allow the reconstruction of how one population or race was derived from another (from Figs. 61 and 62).

and other great apes? If genetic differences are represented as physical distances, we humans are all living in the same mansion—perhaps in adjacent rooms or in rooms in opposite wings of the mansion, but always with the doors open for reproductive purposes. Our closest living evolutionary relatives, the chimpanzees and gorillas, with whom we shared common ancestors millions of years ago, are living in villages on the far horizon (Figure 64).

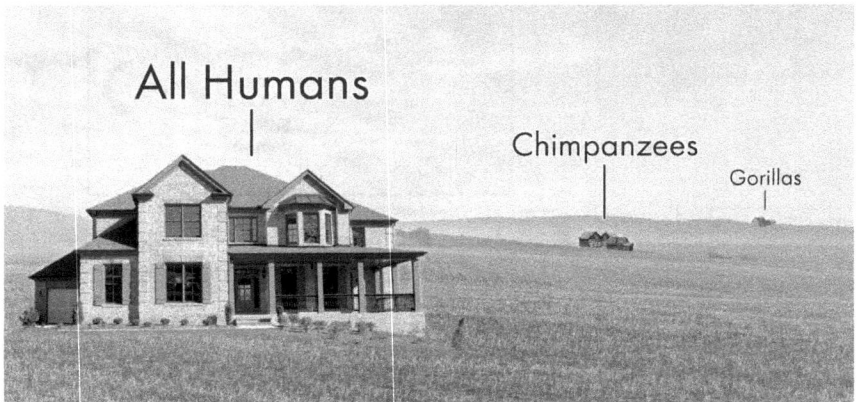

Figure 64. Genetic differences calculated as genetic *distances* between humans and our closest living evolutionary cousins.

Ideally there should be no objection to using the word *race* to refer to any distinguishable human population. An extraterrestrial visitor (named "EV," say) following the usual biological conventions, would not hesitate to use the term to designate the different geographical varieties of humans. EV would not necessarily choose skin color, presence of an eye fold, the shape of the nose or any other physical feature, nor any combination of these to make the distinctions. To tease out the real relations between populations, EV would give up that approach and would objectively measure genetic similarities and differences directly from DNA, using hundreds of thousands of random DNA sequences from hundreds of people all over the globe. From these it would then construct an evolutionary tree of human populations as in Figure 63 above. It is these genetically distinguishable populations, these evolutionary branches, that EV would happily call *races*.

At this stage in her investigation, she might find the idea of "race" useful only in comparing two or maybe three particular populations. Besides, in earthlings' minds, the word sows confusion. Tied up as earthlings are with the word's erroneous implications when applied to their own species, she might recommend giving up the word in most circumstances and using instead the neutral word *population* to designate different human groups.

Racial Prejudice

A tendency to behave antagonistically toward others who are not part of one's own group is unquestionably innate, inherited from distant primate ancestors. Echoes of it can be seen in other modern social primates. The members of a chimpanzee troop know each other and can recognize chimps from other troops, which they will attack and sometimes kill. Baboons recognize who is a member of their own clan and who is not; they stick together with their own clan members and fight with members of other clans for access to food.

While prejudice is innate, as a friend of mine put it, *racial* prejudice is learned. An analogous arrangement exists with human language and language learning. We humans have an innate brain capacity for language; the language we actually adopt is learned. So it is with us humans and prejudice. With regard to other people, we have an innate tendency for prejudicial thinking: *Whom can I trust?—someone like myself, members of my own clan. Whom should I distrust, be wary of, denigrate?—someone different.* The prejudices we actually adopt, the physical appearance of others or their manners of behavior, dress, or speech that we take as significant are learned from our socialization, from parents and relatives and teachers, from experience.

Attitudes and behavior toward others can be readily modified by teaching and learning. In classroom experiments and in psychology laboratories, young people can be led to adopt a discriminatory attitude toward others who are identified just by the color of the clothes they are wearing. In a celebrated classroom exercise carried out by Jane Elliott on her third graders the day after Martin Luther King Jr.'s assassination, some of her students were given (in one description of the experiment) brown fabric collars to wear as a sign of their inferiority. The brown-collared students had to sit in the back of the classroom for the day, were criticized more often for not following rules, were denied special privileges such as second helpings at lunch and extra time at recess, were not allowed to drink at the same water fountain the other children used, and in general were treated so as to make them feel their lower status. By the end of the experiment the collarless students were, all unconsciously, behaving rudely and offensively toward those with the collars, who had become noticeably more timid, less self-assertive, and less proficient in class assignments.

It is unlikely that prejudice specifically based on racial differences would have been built into the human mind from the beginning. Early in human evolution our ancestors probably lived in small tribes. Others they occasionally came in contact with would usually not have been noticeably different from themselves. Long-distance migrations would have occurred only over extended periods of time, maybe many generations, and would have been mostly into regions not previously occupied by any humans at all. What was fundamental was the ability to recognize who was your friend and who was your enemy. (There may have been *species* prejudice between

Homo sapiens and *Homo neanderthalensis* when they encountered each other on intersecting hunting grounds, even though there was some inter-breeding between the two.)

Without significant interracial contacts, early *Homo sapiens* probably operated with an *us-versus-them* mentality, not with *racial* prejudice *per se*. Later, with increasing inter-tribal contacts, discrimination was likely based on different tribal adornments and on different manners of dress and speech. Even later, with increasing interracial contacts, more obvious dissimilarities such as skin color differences probably came into play.

In Summary

Human population differences are a genetic reality and can be measured. The differences are mostly without functional significance. They reflect past spreading of the human species over the globe, although some DNA sequences do reflect evolutionary adaptations to different geographical regions. Such populations (or races, a loose but handy term widely used in some contexts for many plant and animal species) have genetic boundaries between them that are indistinct. They often involve only different frequencies of variable DNA sequences.

Alongside the genetic reality of human population diversity are the social attitudes that people bring to human differences. Such social constructions, which set social rules and behavior, have only tenuous connections with the genetic reality. Social conventions set racial boundaries, boundaries that are neither genetic nor innate.

People cannot help but be defined as belonging to races as socially defined, and the boundaries between them are daily reinforced. These boundaries are more salient than academic investigations of our evolutionary history. Knowing the details about the genetic relations among the different branches of humanity doesn't prevent even people who know better from being impacted by racist ideologies.

An ancient, innate trait found in humans as well as other social primates is the tendency to categorize other members of one's species as friends or enemies—who is in your group and who isn't. With the spread of human populations and more frequent contact between different races, this tendency came to be expressed as racial prejudice. Specifically *racial* prejudice in itself is probably not a fundamental, universal, innate human trait, but a learned attitude.

Latent prejudices can be stimulated by social influences and social experiences. They can be diminished by shared goals and by familiarity with other cultures and distant lands, as when comrades-in-arms meet on foreign battlefields, and in the case of international business conferences, aid organizations, international sports events and—for example—visits of light-skinned scientists from the United States to their dark-skinned counterparts in Jamaica.

19

The Future
of the Only Remaining
Human Species

How sweet I roam'd from field to field,
And tasted all the summer's pride....
—William Blake, *Poetical Sketches* (1783)

The dawn was just beginning to break over the plains of South Dakota, promising a bright clear day. A few birds twittered; a pheasant emitted its squeaky-gate cry and flew up out of the grass toward the nearby woods; farmers slowly awoke planning to begin harvesting their wheat fields. A few cattle nearby and perhaps a few buffalo on the open plains were the first to observe the vivid flash that suddenly brightened the dawn tenfold. Three seconds later the asteroid struck (Figure 65). Millions of tons of vaporized dust and rock exploded up into the atmosphere, and a heat wave engulfed a million square miles of landscape. Five minutes later a thundering shockwave swept over the countryside, flattening everything in its path for a thousand miles. The shock generated earthquakes as far away as Mexico, Alaska, and Japan. For weeks, droplets of melted rock rained down on the earth. Dust circled the globe, blotting out the sun for years and destroying forests all over the world. Plants withered and died. Plant-eating animals starved. Animals that ate the animals that ate plants starved. So, eventually, did humans.

What does the future really hold for our species? There would seem to be five possibilities:

- The human species as we know it will last as long as there is any possibility of life on Earth—perhaps for another billion years or so, toward the end of which time the increasing heat from the expanding Sun will have burnt the Earth to a cinder.

Figure 65. Human extinction by asteroid strike (Bagg Bonanza Farm house photograph by Tallmikejensen4320, Wikimedia Commons, modified).

- *Homo sapiens* will survive but decline as a dominant species on Earth. It will limp along as a minor species for maybe another hundred thousand years, becoming a "living fossil," a species just barely adaptable enough to survive.
- *Homo sapiens* will disappear....
  - ⊚ ...by evolving into some other species. Conditions on Earth will change and *Homo sapiens* will be replaced by an evolutionary descendant, either another species in the genus *Homo*, or a species no longer recognizable as human at all.
  - ⊚ ...by going extinct, without leaving any evolutionary descendants. *Homo sapiens* will vanish from the earth, either as a result of some cataclysmic event such as the one described above, or as a result of a more gradual process such as climate change.

Homo sapiens *Forever*

That our species will last forever, or at least as long as there is life on Earth, is unlikely. Past ages have seen other species of *Homo* and other genera in the hominin line, now all extinct. If the past is any guide to our own uniquely surviving species, *Homo sapiens*, we can ask, how long did other species of *Homo* survive? From the ages of dated fossils (see the table below), the average duration of these other species of *Homo* was about one million years.

Species	Earliest fossil records are from...	Estimated duration of species' existence
Homo sapiens (Africa, now worldwide)	300,000 years ago	?
Homo neanderthalensis (Europe)	400,000 years ago	360,000 years
Homo heidelbergensis (Europe, Africa)	800,000 years ago	600,000 years
Homo ergaster (Africa)	2 million years ago	500,000 years
Homo erectus (Asia)	2.1 million years ago	2 million years
Homo habilis (Africa)	3 million years ago	1.4 million years
Paranthropus robustus (Africa)	2 million years ago	800,000 years
Paranthropus boisei (Africa)	2.5 million years ago	1.2 million years
Australopithecus africanus (Africa)	2.5 million years ago	1 million years
Australopithecus afarensis (Africa)	3.5 million years ago	1 million years
Australopithecus anamensis (Africa)	4 million years ago	400,000 years
Ardipithecus ramidus (Africa)	4 million years ago	1.4 million years
	Average:	**970,000 years**

The average duration of existence for these species is about one million years. For mammals in general, according to a fairly extensive fossil record, the average *mammalian* species also lasted about a million years. So if evolutionary history is any guide, and if we *Homo sapiens* are destined to be an average species of ape on our branch of the evolutionary tree, or even an average mammal, we have lived about a third of our allotted time as a species. In a few hundred thousand years, or less, we will probably follow our ancestors into oblivion.

What drove these ancient apes to extinction? For many, it was probably a change in climate and habitat, and the loss of the plants or animals that constituted the species' diet, combined with insufficient genetic variation to permit rapid evolutionary adaptation—a common reason for species' extinctions generally. This is apparently what happened to the species in the genus *Paranthropus*, who left no evolutionary descendants. In other cases, a subpopulation in the species may have evolved to become a new species, replacing the original species. *Australopithecus anamensis* may have been replaced by *Australopithecus afarensis*, and *Homo heidelbergenesis* may have been replaced by *Homo sapiens* in this way. *Homo sapiens* may have out-competed *Homo neanderthalensis*, the Neanderthals, and driven it to extinction: a case of species competition. This is not a scenario for the demise of *Homo sapiens*, because there is currently no other species of *Homo* competing with us for survival.

Of course, *Homo sapiens* is not an average species. We are exceptional—but whether in ways that will prolong our existence as a species or hasten our disappearance from the Earth is uncertain. Because of our cleverness and our farms and our technology, we might last longer than a million years. Alternatively, because of our success in overpopulating the Earth, depleting our resources, and altering the climate more rapidly than we can adapt to its changes, we might go extinct in a shorter time.

A New Species of Homo?

If our species is not to survive beyond the average time for other species of *Homo*, might our legacy be passed on? *Homo sapiens* might disappear only to be replaced by, say, an evolutionary descendant, *Homo perseverans*.

How does a new species arise from an old one? The evolutionary process commonly proceeds as follows. An old species has been around for a while, sufficiently well adapted to produce enough progeny at each generation, offsetting the death of the individuals of the previous generation. The populations that comprise the species are adapted to their climate, the geography, the trees and other vegetation, the presence of other species upon which they prey and which prey on them. Then something changes. In the geographic range of the species, perhaps over the course of a few centuries or millennia, there is a gradual cooling or warming, more rain, less rain, an earthquake that changes the course of a river, a desert that replaces a forest or vice versa, volcanic eruptions, the elevation of a mountain range, the spreading or migration of a population to an offshore archipelago or to any new location with more or less shade, more or less humidity, closer to or farther from an ocean.

Now a subpopulation of the species may have accumulated some random gene mutations, a few of which happen to improve its adaptation to the new environmental conditions. Under the changed circumstances, because of the advantages the altered genes now begin to give their owners, what were formerly rare genes increase their numbers and become common. Favorable new forms of genes gradually accumulate, making this new population in its new environment different from the populations that made up the old species—different enough that interbreeding between the new and the old populations wanes and eventually vanishes. A new species is born. Under the changed circumstances the new species might out-compete and out-reproduce the old one, which is now less well adapted by comparison. In time, the original species disappears and the new population takes its place. Something like this may have happened as *Homo sapiens* replaced an antecedent species of *Homo*.

Could we *Homo sapiens* in turn lose our place as the dominant and only human species on Earth? Could a future *Homo perseverans* prevail, leaving *Homo sapiens* in the evolutionary dust? We are not likely to be out-competed by any new species of *Homo*. Our species has now spread all over the world, occupying virtually every habitable space. It's true that according to ordinary evolutionary processes operating over tens of thousands of years, different human populations have each adapted physically and physiologically to their own environments, whether it's the savannas of central Africa or the mountains of Tibet or the steppes of Mongolia. The genetic differences underlying these adaptations are not extensive; and the different populations are not sharply separated. As human populations have spread around the globe, the boundaries between them have become blurred, reproduction across them is relatively unconstrained, and we remain one species. There is no island on Earth to which any new subpopulation could retreat and breed among themselves to produce a genetically distinct group of humans. If there were, and an incipient *Homo* species threatened to out-compete and out-reproduce any of our established populations, we would—given what we know about the history of our species and human nature—enslave them or stamp them out with our armies and our economies.

The only uninhabited island paradise that a specialized group of humans might colonize and form a new human species would be another planet or planetary satellite. But Mars, our Moon, and the larger moons of Jupiter and Saturn, are too cold and too dark, and they lack the atmosphere and the resources to support large forms of life for any length of time.

The colonizing humans would have to travel outside our solar system, to other stars in our galaxy with planets circling them. The nearest Star System with a planet at the Goldilocks distance from its star—just right, not too hot, not too cold—is estimated to be about 16 light-years away. Traveling at the speed of light, our human colonists would get there in 16 years. At the more reasonable speed of 30,000 miles per hour (the speed of our fastest space probe), the voyagers would arrive at their destination in about 4000 years. Barring four thousand years of suspended animation, during the voyage they would have to have had not only children, but grandchildren, great-grandchildren, great-great-grandchildren, and so on for at least a hundred generations. As distant in time from the original travelers as we are from the ancient Egyptians, when the descendants finally reached the Goldilocks planet, they might find it uninhabitable because of intense radiation or lack of water or some other reason. The whole enterprise is impractical.

Another way a *Homo perseverans* might replace *Homo sapiens* is by evolution in place. It is not a common means of speciation, but there is evidence for its having occasionally occurred in the history of life. In some *Homo sapiens* population, whatever the environment, new forms of genes would arise giving some individuals an advantage over others in survival or reproduction. The carriers of those genes and their descendants would mix with others, have children with them, and the survival and reproductive success of the whole population would edge upward. With more and more such advantageous mutations spreading farther afield over the generations, like a beneficial virus spreading through the whole population, the species eventually would become different enough from the original that it would in effect be a new species.

What kinds of genes might give significant advantages in survival and reproduction to all who possessed them, so that the genes would likely spread throughout the species? They might be genes that would augment the immune system or boost the body's DNA-repair mechanisms. They might be genes that would make humans even more social, more cooperative and altruistic than they are now, extending people's behavior beyond their current narrow boundaries of nationalism, language, ethnicity, and xenophobia. They might be genes that would enhance the ability to imagine the future and plan for emergencies, overriding selfish short-term concerns, or genes that would diminish the thirst for revenge and prevent it from spilling over reasonable bounds. They might be genes that would enable one to choose a mate who is more faithful and more reliable in matters of child-rearing. Who knows? The possibilities would depend on what mutations happened to occur and their effects on behavior and physiology.

If we allow our imagination to run more wildly, we can dream of the new species an evolution-in-place might produce. As the Earth warms, agriculture, the basis of civilization, withers and dies. Local plants and animals rapidly become depleted. Our species evolves into foragers, developing long thin bodies and even longer and springier legs in order to cover the range necessary for find food. Or, as air pollution increases and the ozone layer is depleted, the surviving humans are those who have retreated to underground tunnels. Their new lifestyle leads to loss of skin pigmentation, the developing of large mole-like front claws for digging, a marked enlargement of the eyes to catch any stray light leaking down into their tunnels, and the development of cockroach-style antennae to help with navigation in the dark.

In another scenario, the Earth warms up, glaciers and ice sheets and the polar ice caps melt, sea levels rise, the continents are inundated, and humans evolve to become a new marine species—a walrus-like creature, but without the tusks, with flippers instead of limbs, and with the nostrils pointing upward. Humans become totally aquatic because there aren't any land or ice floes on which to haul out.

Such poetic visions are not real evolutionary possibilities. Evolution is gradual and slow. It took 25 million years of wallowing through a whole series of species for the present-day walrus body to evolve from its evolutionary ancestor, which was a bear-like land animal. It needed some 15 or 20 million years (and therefore a run of 15 or 20 successive mammalian species) for our distant evolutionary ancestors, small tree-dwelling monkey-like animals, to evolve into our modern human form. Future evolutionary descendants of ours could not achieve a human-to-walrus-like transformation in mere millennia. While it could happen, many successive speciations would be required. By that time the new species would have lost all its humanity, just as the walrus no longer acts like a bear nor thinks like a bear and is in fact no longer a bear.

Besides, environmental change, slow as it appears to us with our short lifespans, is often too rapid for the evolution of most species to keep up. In the whole history of life, the vast majority of species faced with changing environmental conditions did not in fact keep up; they went extinct. The evolutionary tree of life is really an evolutionary bush with hundreds of dead twigs and only a few living sprouts. We are lucky enough to be one of those sprouts, for the time being. It's only natural that we would hold a biased view of our evolutionary history—believing, as we look back, that from the beginning evolution was always aimed straight at us, guaranteeing our survival. But the general rule of extinction will likely apply also to us. We certainly wouldn't be having such confidence in our evolutionary survival if we were one of the extinct ones!

We still harbor most of the genes of our ape-like ancestors. Similarly, in the unlikely event that a *Homo perseverans* does eventually appear and persevere, having evolved from us and having replaced us (a real extinction as far as we are concerned), that new species would carry the many genes it inherited from us, its *Homo sapiens* ancestor. That's a kind of long-lasting life for us, even if only at the gene level. In any case, we are sons and daughters of the Earth, and our line will evolve or go extinct as the Earth changes. Our body form and how long we'll last as *Homo sapiens* are not our choices.

Extinction Once and For All

Several catastrophic events in past eliminated large swaths of plant and animal life. These events were easily large enough to have snuffed out human life as well. Whether a large asteroid strike, the explosion of a nearby star in our galaxy, simultaneous eruptions of several large volcanoes, or significant climate change, humans would likely have succumbed to any such event if we had existed at any one of those times.

The fossil record shows that there have been 10 or 12 major extinctions in the past. The last really big extinction, the one that did in the dinosaurs, occurred 66 million years ago. Therefore, might we expect to suffer another one any millennium now? Such events, however, are ruled by cosmological chance, not by cosmological clockwork. They are rare, separated by intervals of 30 to 80 million years. This is much longer than the one-third of a million years of human existence on Earth, and longer than the usual duration of the average species. We probably don't have to rush to close our doors and windows against the fiery blast any time soon. There may be more imminent doomsday scenarios.

In the Arctic tundra lemmings periodically build up huge populations and after a few years disappear entirely—a population extinguished, apparently. According to legend, large numbers of them commit mass suicide by swarming over the edge of a cliff and plunging into the sea. In our nightmares we can imagine ourselves following their example when the environmental conditions on Earth become intolerable.

As if they have gone locally extinct, whole lemming populations *do* seem to disappear entirely—but only from their usual habitats. What happens is that in winter they deplete all the local beds of moss (their favorite food) and migrate away, sometimes swimming large bodies of water in search of greener pastures. Remnants of the original population survive elsewhere. In three or four years when the moss beds have recovered, the lemmings return to their original homes, and the populations rise again.

Toward the end of its East African existence about one and a half million years ago, one of our distant evolutionary relatives, *Paranthropus boisei*, may have run into the problems of diminishing resources. From studies of ancient dust and pollen samples, animal fossils associated with *Paranthropus*, the jaws and the patterns of tooth wear in the *Paranthropus* fossils themselves, *Paranthropus boisei* probably lived in areas of lakes and open grasslands, subsisting on a diet of grasses, roots, bulbs, seeds, and the leaves and stems of sedges and rushes. Then the climate changed. Over the centuries it rained less often, the lakes dried up, the usual vegetation disappeared, and *Paranthropus boisei* populations declined, finally disappearing like a dwindling thread of a stream dying in the dry African desert. It left no evolutionary descendants.

Restricted as they were to their customary East African habitat, *Paranthropus boisei* populations had nowhere to go to escape the changing climate. Neither do we. *Homo sapiens* is now spread over the whole world, but we have no greener pastures to retreat to while we wait for the planet to convalesce from our misuses of it. The Earth is our only habitat. When we have changed its atmosphere and its oceans and depleted its resources, we can't decamp to another planet and wait for the Earth to recover. We, like *Paranthropus*, will go extinct locally—in our only habitat, our Earth.

Aside from the possibility of our fouling our own Earth-nest, there might be other ways we could go extinct. A viral or bacterial pandemic might decimate the human population, but it wouldn't exterminate the species. The worst epidemic in recorded human history was the 14th century's Black Death, the Great Bubonic Plague. It was caused by bacteria carried by rats and spread by fleas; it killed about one-quarter of the world's population at the time. But the plague didn't reach all parts of the world. Central and Southeast Asia were mostly spared, as was sub–Saharan Africa. And even in Europe half the population survived. The worst viral pandemic in modern times was the 1918–1919 influenza outbreak, which killed about 5 percent of the world's population. Aside from those who escaped because of their high living standards, or by isolating themselves, or by sheer good luck, there must have been some who were naturally resistant, given the great genetic variation that exists in all human populations. We're not like the genetically uniform potatoes that were nearly completely wiped out in Ireland in the mid–1800s by potato blight. Many individuals may die from any disease that afflicts *Homo sapiens*, but the species will survive.

The consequences of a nuclear war might be similar. There would be the deaths from the blasts themselves, from radiation sickness, global radiation fallout, a temporary "nuclear winter" with crop failure, starvation, and social and economic breakdown. The consequences of all these effects have been estimated: in the worst-case scenario, 10 percent of the world's population might be lost. There would be many survivors in non-targeted areas. Life might be miserable, but the species would survive.

Slow Extinction by Climate Change

Because of the current climate change, we *Homo sapiens*, like *Paranthropus* before us, might eventually disappear without leaving any evolutionary descendants.

The present warming trend is likely to continue for decades, if not centuries. It is man-made, the consequence of the increasing amounts of carbon dioxide discharged into the atmosphere ever since the beginning of the Industrial Revolution two and a half centuries ago (Figure 66). While the Sun continues to warm the land and the oceans as usual, the excess CO_2 (and to a lesser extent methane and a fewer other gases) absorbs more of the Sun's direct radiation as well as that reflected back into the atmosphere from the Earth's surface. So the atmosphere heats up even more. The increasing atmospheric CO_2 also acidifies the ocean: $CO_2 + H_2O \rightarrow H_2CO_3$ (carbonic acid).

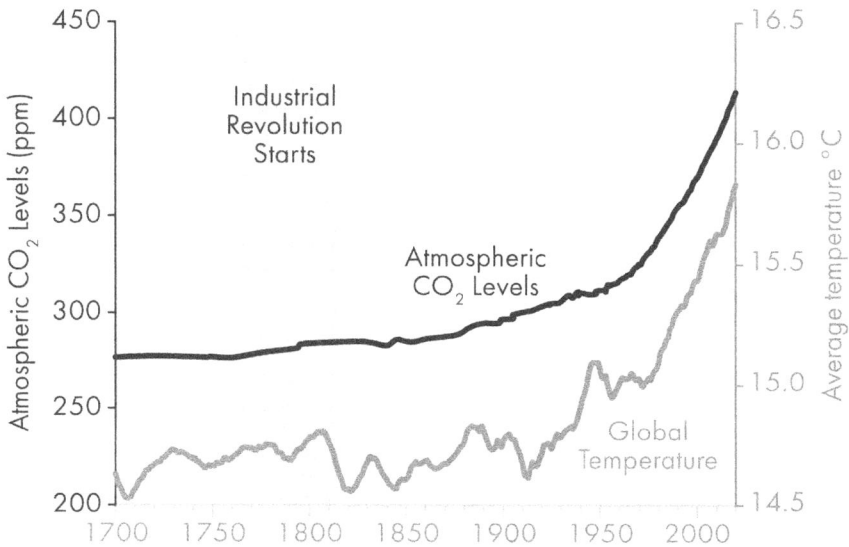

Figure 66. From 1700 to the present, the rise in atmospheric CO_2 levels and the rise in global temperature are correlated (*Temperature Data from years 1 CE–2000 CE*: PAGES 2k Consortium: Neukom, Raphael; Barboza, Luis A.; Erb, Michael P.; Shi, Feng; Emile-Geay, Julien; Evans, Michael N.; et al. [2019]. Reconstruction ensemble median and 95% range. figshare. Dataset. doi. org/10.6084/m9.figshare.8143094.v3. Datasource is full ensemble median which is in units of degrees Celsius anomaly with respect to 1961–1990 CE. *Temperature Data from years 1800 CE–2020 CE*: National Oceanic and Atmospheric Administration at the National Centers for Environmental Information, ncdc.noaa.gov/cag/global/, Global Time Series, Land and Ocean. Datasource is in units of Degrees Celsius Anomaly with respect to 1901–2000 CE. *Data combining*: In order to make the older temperature data have the same units, the average difference between the overlapping 120 years (0.09°C) was added to all the older data. Zero aberration in temperature was set equal to 15°C. Then a 10-previous-year average was taken for the sake of curve-smoothing. *CO_2 Data from years 1006 CE–1978 CE*: Etheridge, D.M.; Steele, L.P.; Langenfelds, R.L.; Francey, R.J.; Barnola, J.M.; Morgan, V.I. [1996]. Natural and anthropogenic changes in atmospheric CO_2 over the last 1000 years from air in Antarctic ice and firn. JGR Atmospheres v101 iD2 p4115–4128 doi. org/10.1029/95JD03410. Data: data.ess-dive.lbl.gov/view/doi:10.3334/CDIAC/ ATG.011. Mean air age used, and multiple values for a given year were averaged. *CO_2 Data from years 1958 CE–2020 CE*: Globally averaged marine surface annual mean data (CSV), Ed Dlugokencky and Pieter Tans, NOAA/GML esrl. noaa.gov/gmd/ccgg/trends/gl_data.html. Monthly data were averaged by year. *Data combining*: For years where the two CO_2 datasets overlapped, an average was used.)

The average global temperature as of the year 2000—land and sea, summer and winter, both hemispheres—is about 15°C (59°F), up from about 13.5°C (56°F) a century ago. That 1.5°C rise in the last century is at least 10 times faster than the estimated rates of temperature increase in any century during the previous millions-of-years periods.

There have been major climate changes in the past history of the Earth. At different times the Earth has been both colder and warmer than it is now. Some of the major episodes of animal and plant extinctions on the planet were correlated with climate changes (Figure 67). Others were probably caused by occasional collisions of asteroids with Earth, by widespread volcanic eruptions and lava flow, or by other geological changes. There haven't been any major periods of extinction that were correlated with climate change for a couple of hundred million years, but we may be on the verge of one now.

Could human technology and know-how save our species, at least for a time, from the wide-ranging effects of climate change? The first step

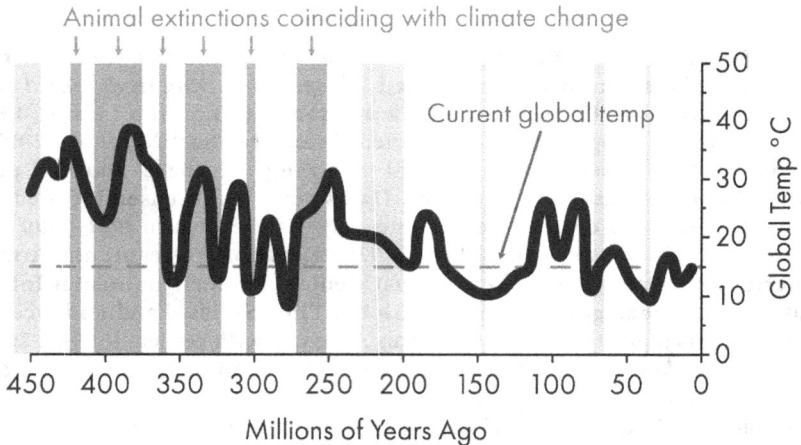

Figure 67. Climate changes in the Earth's history (measured as long-term temperature changes) and periods of major extinctions of animals (vertical shaded bars). Dark vertical bars highlight major extinctions that are correlated with major temperature change; light vertical bars are major extinctions correlated with asteroid strikes, volcanic activity, or other geological changes (graph constructed from data in Raup, C.M., and Sepkoski, J., Jr. 1982. Mass Extinctions in the Marine Fossil Record. *Science* 215:1501–3; Rohde, R.A. and Muller, R.A. 2005. Cycles in fossil diversity. *Nature* 434:208–10; Rampino, M.R. 2018. Dark matter's shadowy effect on Earth. *Astronomy* 46:22–7; Henkes, G.A., et al., 2018. Temperature evolution and the oxygen isotope composition of Phanerozoic oceans from carbonate clumped isotope thermometry. *Earth and Planetary Science Letters* 490:40–50).

would be to reduce the use of CO_2-generating fuels. This could be done by making our way of life more energy efficient, by phasing out fossil fuel use, and by using alternative energy sources: wind, water, and solar power. Capturing solar energy by means of an artificial plant-like photosynthetic system is another possibility. At the same time, means would need to be found to *remove* the already accumulated excess CO_2 from the atmosphere, such as by planting more trees and increasing the area of forests, by reducing our dependence on farm animals, and by employing CO_2-capture-and-storage technologies. Meanwhile, as a stop-gap measure, genetic engineering could produce more heat-tolerant domestic animals and crops.

Humans will have to do more than stop pouring CO_2 into the atmosphere. To reverse the current trend, they will have to begin taking CO_2 *out* of the atmosphere. But even under favorable circumstances, the short-term interests of nations and the public dominate rational policy decisions. Given that the trends in climate change are gradual and long term, the prospects for humanity's willingness to halt or reverse them are dim. Even if the world could take immediate steps to curtail population growth worldwide and to stop generating so much CO_2, the excess CO_2 that has already been dumped into the atmosphere for more than two centuries will endure and will continue to have its effects for thousands of years. Global temperatures will continue to increase, the polar ice caps and glaciers will continue to melt, seas will continue to rise.

As the atmosphere and the ocean continue to warm up, walruses in the north Pacific and north Atlantic are having difficulty finding ice floes on which to haul out and rest and raise their calves. They are retreating to land, the coasts of Siberia and Alaska and Greenland, south of their usual ice floe habitats. Perhaps more and more humans themselves, with continuing sea level rise depriving them of their coasts, might eventually have to turn from land to the ocean for more living space, as imagined in the film *Waterworld*.

Yet the two species—walruses now and humans in the not-too-distant future—face similar predicaments. Walruses, their ice floes disappearing from the northern oceans, find themselves either in the deeper waters of the Arctic Basin, where they are unable to practice their usual habits of scouring shallow sea floors for food, or are forced to spend more time on land, an environment to which they are not adapted. We humans also, as well as our livestock and crops—with our increasing population, the livable land area decreasing, and the equatorial regions of the Earth gradually becoming uninhabitable for greater and greater parts of the year because of extremes of heat and drought—will face increasing rates of starvation and conflict.

Here's what is likely to happen if we stay on our present track. Humans will keep on extracting and burning fossil fuels (currently providing 85 percent of human energy use) with no let-up in sight. Carbon dioxide levels in the atmosphere and in the oceans will continue to rise. The Earth will continue to warm; the climate will continue to change. The oceans will become more acid, coral reefs will continue to die, the physiological and reproductive cycles of fish will continue to be disrupted as fish migrate poleward or lose their spawning grounds altogether. Forests will be more affected by drought and other climate changes than are other land areas and will recover only slowly from climate extremes. More forested land will be claimed for agriculture, livestock production, and human living space. More species will go extinct. Eventually the same factors that are driving non-human species extinct will begin to affect the human species as well: intolerance not of the *average* temperature but of increased frequency of temperature extremes, increased mortality of the very young, loss of living space, decreases in the survival rates of all the domestic animal and plant species upon which humans depend. Droughts will become more common, and the Earth's land surface will become more arid. As humans continue to do what they do best—increase their populations—there will be less food worldwide, less space for farming and agriculture, less living space. Food and energy resources will gradually diminish, and in a few centuries, humans will be put on their own endangered-species list.

The last known human is a scrawny ragged-haired old woman who on August 4, 3121, is tracked down and devoured by a pack of hyenas in eastern Africa—ironically, not far from where, a couple of hundred thousand years ago, some of the first members of our species once lived.

Appendix I

Tissues, Systems,
Organs, and Cell Types

There are 392 cell types listed below. They are organized in four main tissue types (connective, muscle, nervous, epithelial tissue), nine systems, and 30 or 40 different organs (depending on definitions).

Tissue, System, or Organ	*Cell Types*
Connective tissues (support)	
• general	fibroblast, myofibroblast, macrophage, adipocyte, mast cell, mesenchymal cell
• bone	osteoprogenitor cell, osteoblast, osteocyte, osteoclast
• cartilage	chondrocyte, fibroblast
• tendon, ligament	fibroblast
Circulatory system (delivery of oxygen to cells and tissues; carrier of carbon dioxide away from cells and tissues; carrier of nutrient molecules to cells; distribution of hormones to various organs; distribution of cells and molecules of the immune system to all parts of the body)	
• heart	cardiac muscle cell, nerve cell, endothelial cell, fibroblast, adipocyte
• blood & blood cells	erythrocyte, thrombocyte, several kinds of leukocytes (see Immune System)
• blood vessels (arteries, veins, capillaries)	smooth muscle cell, endothelial cell, pericyte, fibroblast
• lymphatic tissue & vessels	several kinds of leukocytes (see Immune System), fibroblast, reticular cell, thymocyte, dendritic cell
Immune System (protection against bacteria, viruses, fungi, parasites)	
• bone marrow	erythroblast, reticulocyte, leukocytes, monoblast, myeloblast, megakaryoblast, megakaryocyte, myeloid stem cell, lymphoid stem cell, adipose cell, adventitial cell, endothelial cell

Tissue, System, or Organ	Cell Types
• white blood cells	neutrophil, eosinophil, basophil, lymphocytes (T-CD4+, T-CD8+, B, NK), plasma cell, monocyte, macrophage, thrombocyte (platelet)
• thymus	lymphatic stem cell, thymocyte, epithelioreticular cells (types I, II, III, IV, V, VI), macrophage, fibroblast
• spleen, tonsils & lymph nodes	fibroblast, myofibroblast, reticular cell, erythrocyte, lymphocytes (B-, T-), macrophage, dendritic cell, leukocytes, endothelial cell

Skin (integument) (protection; temperature regulation; sensation)

• epidermis	basal cell, spinous cell, granular cell, keratinocyte, melanocyte, Langerhans' cell, Merkel's cell
• dermis	fibroblast, smooth muscle cell, adipose cell
• hair & nails	sheath cell, cortical cell, medullary cell, fibroblast
• sebaceous glands	secretory cell, duct cell
• sweat glands	secretory cell, duct cell
• sensory nerves	Schwann cell, fibroblast, supportive cell, endoneurial cell

Respiratory system (exchange of oxygen, carbon dioxide, & water vapor with the environment)

• lungs & bronchioles	alveolar cell type I, alveolar cell type II, Clara cell, goblet cell, ciliated cell, fibroblast, endothelial cell, smooth muscle cell, chondrocyte, alveolar macrophage, exocrine cell
• trachea	basal cell, ciliated columnar cell, mucous cell, brush cell, small-granule cell, tracheal gland cell, fibroblast, chondrocyte, osteocyte

Digestive system (intake, evaluation, & processing of nutrients; secretion of various substances)

• tooth	ameloblast, odontoblast, intermedial cell, stellate reticular cell, outer enamel epithelial cell, cementoblast, cementocyte, fibroblast
• tongue	fibroblast, striated muscle cell, duct cell, serous gland cell, five types of receptor cell, supporting cell, basal cell, epithelial cell
• salivary glands	serous acinar cell, mucous acinar cell, intercalated duct cell, striated duct cell
• esophagus	mucosal gland cell, submucosal gland cell, duct cell, squamous epithelial cell, smooth muscle cell, striated muscle cell, fibroblast
• stomach	gastric gland cell, mucus neck cell, parietal cell, chief cell, enteroendocrine cell, smooth muscle cell, fibroblast
• intestines	enterocyte, smooth muscle cell, fibroblast, goblet cell, Paneth cell, enteroendocrine cell, M cell, adipose cell, lymphocyte, Brunner's gland cell, Auerbach's nerve cell
• liver	hepatocyte, endothelial cell, Kupffer cell, hepatic duct cell

Tissue, System, or Organ	Cell Types
• gallbladder	smooth muscle cell, fibroblast, epithelial cell, mucin-secreting cell, cystic duct cell
• pancreas	acinar cell, centroacinar cell, duct cell
Urinary system (excretion of waste; water & salt balance)	
• kidney	smooth muscle cell, mesangial cell, afferent endothelial cell, efferent endothelial cell, podocyte, parietal cell, PCT cell, juxtaglomerular cell, macula densa cell, epithelial cells of thin segment of loop of Henle (types I, II, III, IV), distal straight tubule cell, DCT cell, collecting duct cell, intercalated cell, fibroblast, macrophage, myofibroblast, lymphatic endothelial cell
• ureter, bladder	urothelial cell, fibroblast, smooth muscle cell
• urethra	urothelial cell, columnar epithelial cell, urethral gland cell, paraurethral gland cell, duct cell
Endocrine system (synthesis & secretion of hormones)	
• pituitary gland	GH cell, PRL cell, ACTH cell, FSH/LH cell, TSH cell, parenchymal cell, basophil, chromophobic cell, neurosecretory cell
• hypothalamus	neurosecretory cells (GHRH/dopamine cell, CRH/somatostatin cell, GNrH/TRH cell)
• thyroid gland	follicular cell, parafollicular cell, endothelial cell
• parathyroid gland	chief cell, oxyphil cell, adipose cell, fibroblast, endothelial cell
• adrenal gland	E-chromaffin cell, NE-chromaffin cell, zona glomerulosa cell, zona fasciculata cell, zona reticularis cell, fibroblast, endothelial cell
• pineal gland	pinealocyte, glial cell
• endocrine cell in testis	Leydig cell
• endocrine cells in pancreas	A-cell, B-cell, D-cell, D-1 cell, PP cell, EC cell
Reproductive system (production of egg and sperm cells, and related functions)	
• ovary	oocyte, follicle cell, stromal cell, granulosa cell, internal thecal cell, external thecal cell, corona radiata cell, granulosa lutein cell, theca lutein cell, atretic follicle cell, granulosa lutein cell, corpus luteum cell, macrophage, neutrophil
• oviduct (Fallopian tube)	mesothelial cell, smooth muscle cell, ciliated cell, peg cell
• uterus	endometrial cell, uterine gland cell, uterine stroma cell, smooth muscle cell, fibroblast, mesothelial cell, cervical gland cell, stratified squamous epithelial cell, columnar epithelial cell, cervical gland cell
• vagina & female genitalia	stratified squamous epithelial cell, basal cell, smooth muscle cell, sebaceous gland cell, sebaceous duct cell

Tissue, System, or Organ	Cell Types
• placenta	cytotrophoblast cell, syncytiotrophoblast, extraembyonic mesodermal cell, chorionic villus cell, decidual cell, endothelial cell
• breast (mammary gland)	lactiferous gland cell, lactiferous duct cell, sebaceous gland cell, sweat gland cell, smooth muscle cell, myoepithelial cell, lymphocyte
• testis	spermatocyte, spermatid, spermatozoon, Leydig cell, Sertoli cell, peritubular contractile cell, rete testis cell, columnar efferent ductule cell, cuboidal efferent ductule cell, principal cells of epididymis, basal cells of epididymis
• seminal vesicle	columnar cell, basal cell, smooth muscle cell
• prostate gland	prostatic secretory cell, basal cell, smooth muscle cell
• bulbourethral gland	bulbourethral gland cell
• urethral gland	urethral gland cell
• penis	endothelial cell, smooth muscle cell, sebaceous gland cell

Nervous system (coordination of body's responses and activity; thinking & feeling)

• brain, spinal cord & nerves	molecular layer cell, external granular layer cell, external pyramidal cell, internal granular layer cell, internal pyramidal cell, fusiform cell, Martinotti cell, polymorphic cell, motor neuron, fibrous astrocyte, protoplasmic astrocyte, Schwann cell, microglial cell, ependymal cell, dura mater cell, pia mater cell, multipolar neuron, sensory neuron, interneuron, satellite cell, perineural cell, basket cell, Purkinje cell

Sense organs (receptors for sensory information)

• eye	scleral cell, corneal epithelial cell, corneal stromal cell, corneal endothelial cell, iris smooth muscle cell, stromal melanocyte of iris, anterior pigmented myoepithelial cell of iris, posterior pigmented epithelial cell of iris, nonpigmented epithelial cell of ciliary process, lens subcapsular epithelial cell, lens fiber cell, retinal ganglion cell, Müller's cell, amacrine cell, bipolar cell, horizontal cell, rod cell, cone cell, retinal pigment epithelial cell, lacrimal gland cell, lacrimal duct cell, Krause accessory gland cell, Wolfring accessory gland cell, sebaceous gland cell of Zeis, Meibomian gland cell, Moll gland cell

Tissue, System, or Organ	Cell Types
• ear	tympanic membrane outer epithelium, tympanic membrane inner epithelium, hair cells (type 1, type II), supporting cells, afferent nerve cell, mesothelial cell, epithelial cell of vestibular membrane, stria vascularis cells (marginal cell, intermediate cell, basal cell), cells of organ of Corti (Böttcher's cell, Claudius' cell, Hensen's cell, outer hair cell, inner hair cell, outer phalangeal cell, inner phalangeal cell, inner pillar cell, outer pillar cell, inner border cell)
• olfactory (smell) receptors	olfactory cell, supporting cell, Bowman's gland cell, Bowman's duct cell
• taste buds	sensory cell, supporting cell, basal cell
• touch (pressure, pain) receptors	Ruffini's corpuscle connective tissue cell, Meissner's corpuscle cells (sensory nerve cell, Schwann cell, supporting cell, capsule cell), Pacinian corpuscle cells (sensory nerve cell, supportive cell), Merkel's cell

Appendix II
Some Musical Preferences

A sampling of personal favorites, examples I consider to be among the highest levels of human musical art, some of which still bring tears to my eyes.

Performance

- Sandra Hurtado-Ros, singing a Bernart de Ventadorn troubadour song
- Maria Callas, operatic arias
- Sandra Sangiao with the Barcelona Gipsy balKan Orchestra, especially the concerts in Andorra in 2015, and the last concert in the Sala Apolo, Barcelona, in 2019
- David Russell's guitar concerts
- Ana Vidović on classical guitar, playing "La Catedral" by Agustín Barrios Mangoré
- Pavlo, *Live in Kastoria*
- Anna Fedorova, *The Rachmaninoff Piano Concertos* and *Rachmaninoff's Rhapsody on a Theme of Paganini*
- Martha Argerich, *Maurice Ravel's Piano Concerto in G Major*
- Alice Sara Ott, *Beethoven's Piano concerto No. 3*, Tchaikovsky's *First Piano Concerto*, Grieg's *Piano Concerto in A minor, Op. 16*, Mussorgsky's *Pictures at an Exhibition*
- Yo-Yo Ma, Stuart Duncan, Edgar Meyer, Chris Thile, *The Goat Rodeo Sessions*
- Khatia Buniatishvili, *Rachmaninoff's Piano Concerto No. 3*

Composition

- Luigi Boccherini, *Sonatas for Violin and Harpsichord*
- Gabriel Fauré, *Requiem*
- Freddy Hubbard, *Straight Life* and *First Light*
- Astor Piazzolla, *Adiós Nonino* and *María de Buenos Aires*
- Sergei Rachmaninoff, *Piano Concerto No. 2 in C minor, Op. 18*
- Maurice Ravel, *La Valse*
- Franz Schubert, *String Quartet No. 14 in D minor (Death and the Maiden)*
- Jean Sibelius, *The Swan of Tuonela*

Background
and Further Reading

Introduction

Antón, S.C., Potts, R., and Aiello, L.C. 2014. Evolution of early homo: An integrated biological perspective. *Science* 345:1236828.

Betts, H.C., et al. 2018. Integrated genomic and fossil evidence illuminates life's early evolution and eukaryote origins. *Nature Ecology and Evolution* 2:1556–62.

Cartmill, M. 2012. Primate origins, human origins, and the end of higher taxa. *Evolutionary Anthropology* 21:208–20.

Dawkins, R. 1995. *River Out of Eden: A Darwinian View of Life*. New York: Basic Books.

Ebersberger, I., et al. 2007. Mapping human genetic ancestry. *Molecular Biology and Evolution* 24:2266–76.

Foley, R.A. 2016. Mosaic evolution and the pattern of transitions in the hominin lineage. *Philosophical Transactions of the Royal Society of London, Series B, Biological Sciences* 371:20150244.

Margulis, L., and Schwartz, K.V. 1997. *Five Kingdoms: An Illustrated Guide to the Phyla of Life on Earth* (3rd ed.). New York: W.H. Freeman, Macmillan.

McHenry, H.M. 1994. Tempo and mode in human evolution. *Proceedings of the National Academy of Sciences USA* 91:6780–86.

Rougier, H., et al. 2007. Peştera cu Oase 2 and the cranial morphology of early modern Europeans. *Proceedings of the National Academy of Sciences USA* 104:1165–70.

Schreiber, U., and Mayer, C. 2020. *The First Cell*. New York: Springer.

Chapter 1

Robots

Blair, A., and Saffidine, A. 2019. AI surpasses humans at six-player poker. *Science* 365 864–65.

Crick, C., and Scassellati, B. 2010. Controlling a robot with intention derived from motion. *Topics in Cognitive Science* 2:114–26.

Dumouchel, P., and Damiano, L. 2017. *Living with Robots*. Cambridge: Harvard University Press.

Konidaris, G., Kaelbling, L.P., and Lozano-Perez, T. 2018. From skills to symbols: Learning symbolic representations for abstract high-level planning. *Journal of Artificial Intelligence Research* 61:215–89.

Science. 2014, 10 October. *Special Issue: The Social Life of Robots* 346(6206).

Wallach, W., and Allen, C. 2008. *Moral Machines: Teaching Robots Right from Wrong*. Oxford: Oxford University Press.

Philosophy

Blackmore, S., and Troscianko, E.T. 2018. *Consciousness: An Introduction* (3rd ed.). Abingdon-on-Thames: Routledge.

Cairns-Smith, A.G. 1996. *Evolving the Mind: On the Nature of Matter and the Origin of Consciousness*. Cambridge: Cambridge University Press.

Custers, R., and Aarts, H. 2010. The unconscious will: how the pursuit of goals operates outside of conscious awareness. *Science* 329:47–50.

Dennett, D.C. 2009. The part of cognitive science that is philosophy. *Topics in Cognitive Science* 1:231–36.

Gray, J. 2015. *The Soul of the Marionette: A Short Inquiry into Human Freedom*. New York: Farrar, Straus & Giroux.

Ryle, G. 1949. *The Concept of Mind*. London: Hutchinson.

Searle, J.R. 2009. *Mind: A Brief Introduction*. Oxford: Oxford University Press.

Chapter 2

Bustamante, C.D., et al. 2005. Natural selection on protein-coding genes in the human genome. *Nature* 437:1153–57.

Iwasa, J., and Marshall, W. 2020. *Karp's Cell and Molecular Biology* (9th ed.). Hoboken: Wiley.

Lewis, R. 2018. *Human Genetics: Concepts and Applications* (12th ed.). New York: McGraw-Hill.

Margulis, L., and Sagan, D. 1995. *What Is Life?* New York: Simon & Schuster.

Nurse, P. 2000. The incredible life and times of biological cells. *Science* 289:1711–16.

Thomas, L. 1975. *The Lives of a Cell*. London: Penguin.

Chapter 3

Chazaud, C., and Yamanaka, Y. 2016. Lineage specification in the mouse preimplantation embryo. *Development* 143:1063–74.

De Paepe, C., et al. 2014. Totipotency and lineage segregation in the human embryo. *Molecular Human Reproduction* 20:599–618.

Edwards, R.G. 2000. The role of embryonic polarities in preimplantation growth and implantation of mammalian embryos. *Human Reproduction*, Suppl 6:1–8.

Hansis, C., and Edwards, R.G. 2003. Cell differentiation in the preimplantation human embryo. *Reproductive Biomedicine Online* 6:215–20.

Zernicka-Goetz, M., and Highfield, R. 2020. *The Dance of Life: The New Science of How a Single Cell Becomes a Human Being*. New York: Basic Books.

Chapter 4

Balter, M. 2005. Are humans still evolving? *Science* 309:234–37.

Coyne, J. 2009. *Why Evolution Is True*. New York: Viking Penguin.

Dawkins, R. 1995. *River Out of Eden*. New York: Basic Books.

Dawkins, R. 1997. *Climbing Mt. Improbable*. London: W.W. Norton.

Dawkins, R. 1998. *Unweaving the Rainbow: Science, Delusion and the Appetite for Wonder*. New York: Houghton Mifflin Harcourt.

Dennett, D.C. 1995. *Darwin's Dangerous Idea: Evolution and the Meanings of Life*. New York: Simon & Schuster.

Fan, S., Hansen, M.E., Lo, Y., and Tishkoff, S.A. 2016. Going global by adapting local: A review of recent human adaptation. *Science* 354:54–59.

Krimsky, S., and Gruber, J., eds. 2013. *Genetic Explanations: Sense and Nonsense*. Cambridge: Harvard University Press.

Newman, S.A. 2008. Evolution: The public's problem, and the scientists.' *Capitalism, Nature, Socialism* 19(1):98–106.

Prothero, D.R. 2020. *The Story of Evolution in 25 Discoveries: The Evidence and the People Who Found It*. New York: Columbia University Press.

Chapter 5

Ayala, F.J., and Cela-Conde, C.J. 2017. *Processes in Human Evolution: The Journey from Early Hominins to Neanderthals and Modern Humans* (2nd ed.). Oxford: Oxford University Press.

Coyne, J.A. 2009. *Why Evolution Is True*. New York: Penguin.

Dawkins, R. 1995. *River Out of Eden: A Darwinian View of Life*. New York: Basic Books.

Dawkins, R. 1998. *Unweaving the Rainbow: Science, Delusion and the Appetite for Wonder*. New York: Houghton Mifflin Harcourt.

Dawkins, R. 2016. *Climbing Mount Improbable*. New York: W.W. Norton.

Dennett, D.C. 1995. *Darwin's Dangerous Idea: Evolution and the Meanings of Life*. New York: Simon & Schuster.

Maynard Smith, J., and Szathmáry, E. 1995. *The Major Transitions in Evolution*. Oxford: Oxford University Press.

Mohun, J., Winston, R., Wilson, D.E., et al., eds. 2006. *Human*. London: Doring Kindersley and New York: DK.

Prothero, D.R. 2017. *Evolution: What the Fossils Say and Why It Matters* (2nd ed.). New York: Columbia University Press.

Chapter 6

Burness, G.P., Diamond, J., and Flannery, T. 2001. Dinosaurs, dragons, and dwarfs: The evolution of maximal body size. *Proceedings of the National Academy of Sciences USA* 98:14518–23.

Herridge, V.L., and Lister, A.M. 2012. Extreme insular dwarfism evolved in a mammoth. *Proceedings of the Royal Society B, Biological Sciences* 280:3193–200.

Libby, E. and Ratcliff, W.C. 2014. Ratcheting the evolution of multicellularity. *Science* 346:426–27.

Niklas, K.J., and Newman, S.A. 2013. The origins of multicellular organisms. *Evolution and Development* 15:41–52.

Okie J.G., et al. 2013. Effects of allometry, productivity and lifestyle on rates and limits of body size evolution. *Proceedings of the Royal Society B, Biological Sciences* 280:20131007.

Pilbeam, D., and Gould, S.J. 1974. Size and scaling in human evolution. *Science* 186:892–901.

Chapter 7

Cooke, J. 2004. Developmental mechanism and evolutionary origin of vertebrate left/right asymmetries. *Biological Reviews, Cambridge Philosophical Society* 79:377–407.

Fedonkin, M.A., and. Waggoner, B.M. 1997. The Late Precambrian fossil *Kimberella* is a mollusc-like bilaterian organism. *Nature* 388:868–71.

Knoll, A.H. and Carroll, S.B. 1999. Early animal evolution: Emerging views from comparative biology and geology. *Science* 284:2129–37.

Pecoits E., et al. 2012. Bilaterian burrows and grazing behavior at >585 million years ago. *Science* 336:1693–96.

Chapter 8

Bell, G. 1988. *Sex and Death in Protozoa: The History of Obsession.* Cambridge: Cambridge University Press.

Bellis, M.A., Hughes, K., Hughes, S., and Ashton, J.R. 2005. Measuring paternal discrepancy and its public health consequences. *Journal of Epidemiology and Community Health* 59:749–54.

Candolin, U., and Vlieger, L. 2013. Should attractive males sneak: The trade-off between current and future offspring. *PLoS One* 8:e57992.

Cohas, A., and Allainé, D. 2009. Social structure influences extra-pair paternity in socially monogamous mammals. *Biology Letters* 5:313–16.

Lovejoy, C.O. 2009. Reexamining Human Origins in Light of *Ardipithecus ramidus. Science* 326:74, 74e1–74e8.

Maynard Smith, J. 1978. *The Evolution of Sex.* Cambridge: Cambridge University Press.

Michod, R.E. 1995. *Eros and Evolution: a Natural Philosophy of Sex.* Boston: Addison-Wesley.

Chapter 9

Blackburn, D.G., and Sidor, C.A. 2014. Evolution of viviparous reproduction in Paleozoic and Mesozoic reptiles. *The International Journal of Developmental Biology* 58:935–48.

Bounous, G., Kongshavn, P.A., Taveroff, A., and Gold, P. 1988. Evolutionary traits in human milk proteins. *Medical Hypotheses* 27:133–40.

Hinde, K., and Milligan, L.A. 2011. Primate milk: Proximate mechanisms and ultimate perspectives. *Evolutionary Anthropology* 20:9–23.

Kawasaki, K., Lafont, A.G., and Sire, J.Y. 2011. The evolution of milk casein genes from tooth genes before the origin of mammals. *Molecular Biology and Evolution* 28:2053–61.

Kunz, T.H. and Hosken, D.J. 2009. Male lactation: Why, why not and is it care? *Trends in Ecology and Evolution* 24:80–85.

Oftedal, O.T. 2002. The mammary gland and its origin during synapsid evolution. *Journal of Mammary Gland Biology and Neoplasia* 7:225–52.

Oftedal O.T. 2002. The origin of lactation as a water source for parchment-shelled eggs. *Journal of Mammary Gland Biology and Neoplasia* 7:253–66.

Oftedal O.T. 2012. The evolution of milk secretion and its ancient origins. *Animal* 6:355–68.

Panciroli, E. 2021. *Beasts Before Us: The Untold Story of Mammal Origins and Evolution.* New York: Bloomsbury Sigma.

Quinlan, R.J., and Quinlan, M.B. 2008. Human lactation, pair-bonds, and alloparents: A cross-cultural analysis. *Human Nature* 19:87–102.

Chapter 10

d'Errico, F., et al. 2012. Early evidence of San material culture represented by organic artifacts from Border Cave, South Africa. *Proceedings of the National Academy of Sciences USA* 109:13214–19.

Gilligan, I. 2010. The prehistoric development of clothing: archeological implications of a thermal model. *Journal of Archeological Method and Theory* 17:15–80.

Kamberov, Y.G., et al. 2018. Comparative evidence for the independent evolution of hair and sweat gland traits in primates. *Journal of Human Evolution* 125:99–105.

Li, W., et al. 2010. Genotyping of human lice suggests multiple emergencies of body lice from local head louse populations. *PLoS Neglected Tropical Diseases* 4:e641.

Perry, G.H. 2014. Parasites and human evolution. *Evolutionary Anthropology* 23:218–28.

Ruxton, G.D., and Wilkinson, D.M. 2011. Avoidance of overheating and selection for both hair loss and bipedality in hominins. *Proceedings of the National Academy of Sciences USA* 108:20965–69.

Toups, M.A., Kitchen, A., Light, J.E., and Reed, D.L. 2011. Origin of clothing lice indicates early clothing use by anatomically modern humans in Africa. *Molecular Biology and Evolution* 28:29–32.

Yesudian, P. 2011. Human hair—an evolutionary relic? *International Journal of Trichology* 3:69.

Chapter 11

Crompton, A.W., Taylor, C.R., and Jagger, J.A. 1978. Evolution of homeothermy in mammals. *Nature* 272:333–36.

Grigg, G.C., Beard, L.A. and Augee, M.L. 2004. The evolution of endothermy and its diversity in mammals and birds. *Physiological and Biochemical Zoology* 77:982–97.

Legendre, L.J., and Davesne, D. 2020. The evolution of mechanisms involved in vertebrate endothermy. *Philosophical Transactions of the Royal Society of London, Series B, Biological Sciences* 375:20190136.

Lovegrove, B.G. 2017. A phenology of the evolution of endothermy in birds and mammals. *Biological Reviews of the Cambridge Philosophical Society* 92:1213–40.

Rowland, L.A., Bal, N.C., and Periasamy, M. 2015. The role of skeletal-muscle-based thermogenic mechanisms in vertebrate endothermy. *Biological Reviews of the Cambridge Philosophical Society* 90:1279–97.

Tansey, E.A., and Johnson, C.D. 2015. Recent advances in thermoregulation. *Advances in Physiology Education* 39:139–48.

Chapter 12

Bramble, D.M., and Lieberman, D.E. 2004. Endurance running and the evolution of Homo. *Nature* 432:345–52.

DeSilva, J.M., McNutt, E.J., Benoit, J., and Zipfel, B. 2019. One small step: A review of Plio-Pleistocene hominin foot evolution. *American Journal of Physical Anthropology* 168, Suppl 67:63–140.

Gruss, L.T., and Schmitt, D. 2015. The evolution of the human pelvis: Changing adaptations to bipedalism, obstetrics and thermoregulation. *Philosophical Transactions of the Royal Society of London, Series B, Biological Sciences* 370:20140063.

Holowka, N.B., and Lieberman, D.E. 2018. Rethinking the evolution of the human foot: Insights from experimental research. *Journal of Experimental Biology* 6:221(Pt. 17).

McNutt, E.J., Zipfel, B., and DeSilva, J.M. 2018. The evolution of the human foot. *Evolutionary Anthropology* 27:197–217.

Pontzer, H. 2017. Economy and endurance in human evolution. *Current Biology* 27(12):R613-R621.

Chapter 13

Chen, J., Zou, Y., Sun, Y.-H., and Ten Cate, C. 2019. Problem-solving males become more attractive to female budgerigars. *Science* 363:166–7

Crick, F. 1995. *The Astonishing Hypothesis: The Scientific Search for the Soul*. New York: Scribner.

Dunbar, R.I.M., and Shultz, S. 2017. Why are there so many explanations for primate brain evolution? *Philosophical Transactions of the Royal Society of London, Series B, Biological Sciences* 372:pii:20160244.

Herculano-Houzel, S. 2012. The remarkable, yet not extraordinary, human brain as a scaled-up primate brain and its associated cost. *Proceedings of the National Academy of Sciences USA* 109, Suppl 1:10661–68.

Herculano-Houzel, S. 2016. *The Human Advantage: A New Understanding of How Our Brains Became Remarkable*. Cambridge: MIT Press.

Herculano-Houzel, S. 2019. Life history changes accompany increased numbers of cortical neurons: A new framework for understanding human brain evolution. *Progress in Brain Research* 250:179–216.

Hill, K., and Boyd, R. 2021. Behavioral convergence in humans and animals. *Science* 271:235–36.

Hill, K., et al., 2009. The emergence of human uniqueness: Characters underlying behavioral modernity. *Evolutionary Anthropology* 18:187–200.

Kaas J.H. 2013. The evolution of brains from early mammals to humans. *Wiley Interdisciplinary Reviews. Cognitive Science* 4:33–45.

McBrearty, S., and Brooks, A. 2000. The revolution that wasn't: A new interpretation of the origin of modern human behavior. *Journal of Human Evolution* 39:453–563.

Mellars, P. 2005. The impossible coincidence: A single-species model for the origins of modern human behavior in Europe. *Evolutionary Anthropology* 14:12–27.

Olkowicz, S., et al. 2016. Birds have primate-like numbers of neurons in the forebrain. *Proceedings of the National Academy of Sciences USA* 113: 7255–60.

Parmigiani, S., Pievani, T., and Tattersall, I. 2016. What made us human? Biological and cultural evolution of *Homo sapiens*. *Journal of Anthropological Sciences* 94:1–4.

Powell A., Shennan S., and Thomas, M.G. 2009. Late Pleistocene demography and the appearance of modern human behavior. *Science* 324:1298–301.

Roth, G., and Dicke, U. 2012. Evolution of the brain and intelligence in primates. *Progress in Brain Research* 195:413–30.

Stout, D. 2011. Stone toolmaking and the evolution of human culture and cognition. *Philosophical Transactions of the Royal Society of London, Series B, Biological Sciences* 366:1050–59.

Tattersall, I. 2009. Human origins: Out of Africa. *Proceedings of the National Academy of Sciences USA* 106:16018–21.

Tattersall, I. 2010. Human evolution and cognition. *Theory in Biosciences* 129:193–201.

Chapter 14

Barash, D.P., and Lipton, J.E. 2009. *How Women Got Their Curves, and Other Just-So Stories*. New York: Columbia University Press.

Brassey, C.A., Gardiner, J.D., and Kitchener, A.C. 2018. Testing hypotheses for the function of the carnivoran baculum using finite-element analysis. *Philosophical Transactions of the Royal Society of London, Series B, Biological Sciences* 285:20181473.

de Waal, F. 2005. *Our Inner Ape: A Leading Primatologist Explains Why We Are Who We Are*. Ch. 3 ("Sex"). New York: Penguin.

Diamond, J. 1997. *Why Is Sex Fun? The Evolution of Human Sexuality*. New York: Basic Books.

Dixon, A. 2012. *Primate Sexuality: Comparative Studies of the Prosimians, Monkeys, Apes, and Humans* (2nd ed.). New York: Oxford University Press.

Fürtbauer, I., Heistermann, M., Schülke, O., and Ostner, J. 2011. Concealed fertility and extended female sexuality in a non-human primate (*Macaca assamensis*). *PLoS One* 6:e23105.

Hill, K., and Hurtado, A.M. 2009. Cooperative breeding in South American hunter-gatherers. *Proceedings of the Royal Society, Series B, Biological Sciences* 276: 3863–70.

Lovejoy, C.O. 2009. Reexamining human origins in light of *Ardipithecus ramidus*. *Science* 326:74, 74e1–74e8.

Martin R.D. 2007. The evolution of human reproduction: a primatological perspective. *American Journal of Physical Anthropology* Suppl 45:59–84.

Michod, R.E. 1996. *Eros and Evolution: a Natural Philosophy of Sex*. New York: Basic Books.

Pfaus, J.G., Scardochio, T., Parada, M., Gerson, C., Quintana, G.R., and Coria-Avila, G.A. 2016. Do rats have orgasms? *Socioaffective Neuroscience and Psychology* 6:31883.

Possamai, C.B., Young, R.J., Mendes, S.L., and Strier, K.B. 2007. Socio-sexual behavior of female northern muriquis (*Brachyteles hypoxanthus*). *American Journal of Primatology* 69:766–76.

Shih, C.-K. 2009. *Quest for Harmony: The Moso Traditions of Sexual Union and Family Life.* Stanford: Stanford University Press.

Sillen-Tullberg, B., and Moller, A.P. 1993. The relationship between concealed ovulation and mating systems in anthropoid primates: A phylogenetic analysis. *American Naturalist* 141:1–25.

Thornhill, R., and Gangestad, S.W. 2008. *The Evolutionary Biology of Human Female Sexuality.* New York: Oxford University Press.

van Schaik, C. 2004. *Among Orangutans: Red Apes and the Rise of Human Culture.* Cambridge: Belknap Press (Harvard University Press).

Walker, R.S., Flinn, M.V., and Hill, K.R. 2010. Evolutionary history of partible paternity in lowland South America. *Proceedings of the National Academy of Sciences USA* 107:19195–200.

Willems, E.P., and van Schaik, C.P. 2017. The social organization of *Homo ergaster*: Inferences from anti-predator responses in extant primates. *Journal of Human Evolution* 109:11–21.

Zuk, M. 2002. *Sexual Selections: What We Can and Can't Learn About Sex from Animals.* Berkeley: University of California Press.

Chapter 15

Batson, C.D. 1990. How social an animal? The human capacity for caring. *American Psychologist* 45:336–46.

de Waal, F. 2005. *Our Inner Ape: A Leading Primatologist Explains Why We Are Who We Are.* New York: Berkley.

de Waal, F., ed. 2002. *Tree of Origin: What Primate Behavior Can Tell Us about Human Social Evolution.* Cambridge: Harvard University Press.

Griffin, D.R. 2001. *Animal Minds: Beyond Cognition to Consciousness.* Chicago: University of Chicago Press.

Hare, B., and Woods, V. 2020. *Survival of the Friendliest: Understanding Our Origins and Rediscovering Our Common Humanity.* New York: Random House.

Hublin, J.-J. 2009. The prehistory of compassion. *Proceeding of the National Academy of Sciences USA* 106:6429–30.

Kelly, R.L. 2013. *The Lifeways of Hunter-Gatherers: The Foraging Spectrum* (2nd ed.). Cambridge: Cambridge University Press.

Martin, M.K. 2019. *Social DNA: Rethinking Our Evolutionary Past.* New York: Berghahn Books.

Moffett, M.W. 2019. *The Human Swarm: How Our Societies Arise, Thrive, and Fall.* New York: Basic Books.

Silk, J.B. 2011. Evolutionary biology: The path to sociality. *Nature* 479:182–83.

Tooby, J., and Cosmides, L. 2015. "Conceptual Foundations of Evolutionary Psychology," vol. 1, Chapter 1, in David M. Buss, ed., *The Handbook of Evolutionary Psychology* (2nd ed.). New York: John Wiley & Sons.

Torréns, M.G., and Kärtner, J. 2017. The influence of socialization on early helping from a cross-cultural perspective. *Journal of Cross-Cultural Psychology* 48:353–68.

Ward, E., Ganis, G., and Bach, P. 2019. Spontaneous vicarious perception of the content of another's visual perspective. *Current Biology* 29:874–80.

Warneken, F., and Tomasello, M. 2009. Varieties of altruism in children and chimpanzees. *Trends in Cognitive Sciences* 13:397–402.

Chapter 16

Ellenberger, C. 2018. *Theme and Variations: Musical Notes by a Neurologist.* Palm Springs: Sunacumen Press.

Fenk-Oczlon, G., and Fenk, A. 2009. Some parallels between language and music from a cognitive and evolutionary perspective. *Musicae Scientiae* 13 (issue 2, suppl.):201–26.

Fitch, W.T. 2016. Dance, music, meter and groove: A forgotten partnership. *Frontiers in Human Neuroscience* 10:64.

Honing, H. 2018, March 15. On the biological basis of musicality. *Annals of the New York Academy of Sciences*, doi: 10.1111/nyas.

Honing, H., Bouwer, F.L., Prado, L., and Merchant, H. 2018. Rhesus Monkeys (*Macaca mulatta*) sense isochrony in rhythm, but not the beat: Additional support for the gradual audiomotor evolution hypothesis. *Frontiers in Neuroscience* 12:475.

Killin, A. 2017. Plio-Pleistocene foundations of Hominin musicality: Coevolution of cognition, sociality, and music. *Biological Theory* 12:222–35.

Masataka, N. 2009. The origins of language and the evolution of music: A comparative perspective. *Physics of Life Reviews* 6:11–22.

Mithen, S. 2006. *The Singing Neanderthals: The Origins of Music, Language, Mind and Body*. Cambridge: Harvard University Press.

Richter J., and Ostovar, R. 2016. "It don't mean a thing if it ain't got that swing": An alternative concept for understanding the evolution of dance and music in human beings. *Frontiers in Human Neuroscience* 10:485.

Trehub, S.E. 2001. Musical predispositions in infancy. *Annals of the New York Academy of Sciences* 930:1–16.

Chapter 17

Botha, R., and Knight, C., eds. 2009. *The Cradle of Language (Oxford Studies in the Evolution of Language)*. New York: Oxford University Press.

Engesser, S., Ridley, A.R., and Townsend, S.W. 2016. Meaningful call combinations and compositional processing in the southern pied babbler. *Proceedings of the National Academy of Sciences USA* 113:5976–81.

Fischer, J., 2017. Primate vocal production and the riddle of language evolution. *Psychonomic Bulletin & Review* 24:72–78.

Fitch, W.T., de Boer, B., Mathur, N., and Ghazanfar, A.A. 2016. Monkey vocal tracts are speech-ready. *Science Advances* 2:e1600723.

Hauser, M.D., et al. 2014. The mystery of language evolution. *Frontiers in Psychology* 5: Article 401.

Hill, K., Barton, M., and Hurtado, M. 2009. The emergence of human uniqueness: characters underlying behavioral modernity. *Evolutionary Anthropology* 18:187–200.

Hurford, J.R. 2014. *The Origins of Language: A Slim Guide*. Oxford: Oxford University Press.

Kenneally, C. 2007. *The First Word: The Search for the Origins of Language*. New York: Viking.

McBrearty, S., and Brooks, A. 2000. The revolution that wasn't: A new interpretation of the origin of modern human behavior. *Journal of Human Evolution* 39:453–563.

Nowak, M.A., and Komarova, N.L. 2001. Towards an evolutionary theory of language. *Trends in Cognitive Sciences* 5:288–95.

Pepperberg, I. 2009. *Alex & Me: How a Scientist and a Parrot Discovered a Hidden World of Animal Intelligence—and Formed a Deep Bond in the Process*. New York: Harper Perennial.

Pinker, S. 2007. *The Language Instinct: How the Mind Creates Language*. New York: HarperCollins.

Rousseau, J.-J., and Herder, J.G. 1852, 1891. *On the Origin of Language*. Reprint edition 1986. Chicago: University of Chicago Press.

Chapter 18

Allport, G.W. 1979. *The Nature of Prejudice* (25th anniv. Ed.). New York: Addison-Wesley.

American Psychological Association. 2005. Genes, race, and psychology in the genome era. *American Psychologist*, Special issue 60(1):5–103.

Douka, K., O'Reilly, M., and Petraglia, M.D. 2017. On the origin of modern humans: Asian perspectives. *Science* 358: eaai9067.

Fan, S., Hansen, M.E., Lo, Y., and Tishkoff, S.A. 2016. Going global by adapting local: A review of recent human adaptation. *Science* 354:54–59.

Lahn, B.T., and Ebenstein, L. 2009. Let's celebrate human genetic diversity. *Nature* 461:726–28.

Lewontin, R. 1995. *Human Diversity*. New York: W.H. Freeman.

Long, J.C., Li, J., and Healy, M.E. 2009. Human DNA sequences: More variation and less race. *American Journal of Physical Anthropology* 139:23–34.

Moffett, M.W. 2019. *The Human Swarm: How Our Societies Arise, Thrive, and Fall*. New York: Basic Books.

Molnar, S. 2017. *Human Variation: Races, Types, and Ethnic Groups* (6th ed.). Abingdon-on-Thames: Routledge.

Nielsen, R., Akey, J.M., Jakobsson, M., Pritchard, J.K., Tishkoff, S., and Willerslev, E. 2017. Tracing the peopling of the world through genomics. *Nature* 541:302–10.

Reich, D. 2018. *Who We Are and How We Got Here: Ancient DNA and the New Science of the Human Past*. New York: Vintage.

Chapter 19

Brannen, P. 2017. *The Ends of the World: Volcanic Apocalypses, Lethal Oceans, and Our Quest to Understand Earth's Past Mass Extinctions*. New York: HarperCollins.

Calleja-Agius, J., England, K., and Calleja, N. 2020, October 6. The effect of global warming on mortality. *Early Human Development* 105222.

Gilding, P. 2019. Report: What is a climate emergency and does the evidence justify one? *Breakthrough: National Centre for Climate Restoration* (breakthroughonline.org.au).

Henkes, G.A., et al. 2018. Temperature evolution and the oxygen isotope composition of Phanerozoic oceans from carbonate clumped isotope thermometry. *Earth and Planetary Science Letters* 490:40–50.

Leslie, J. 1996. *The End of the World: The Science and Ethics of Human Extinction*. Abingdon-on-Thames: Routledge.

Martin, B. 1982. Critique of nuclear extinction. *Journal of Peace Research* 19:287–300.

Matheny, J.G. 2007. Reducing the risk of human extinction. *Risk Analysis* 27:1335–44.

McMichael, A. 2017. *Climate Change and the Health of Nations: Famines, Fevers, and the Fate of Populations*. Oxford: Oxford University Press.

Rau, G.H., and Lackner, K.S. 2013. Reversing excess atmospheric CO_2. *Science* 340:1522–23. Response: Matthews, D., and Solomon, S. 340:1523.

Raup, D.M. 1986. Biological extinction in earth history. *Science* 231:1528–33.

Rich, N. 2019. *Losing Earth: A Recent History*. New York: Farrar, Straus and Giroux.

Román-Palacios, C., and Wiens, J.J. 2020. Recent responses to climate change reveal the drivers of species extinction and survival. *Proceedings of the National Academy of Sciences USA* 117:4211–17.

Shaw, J. 2020. Controlling the global thermostat. *Harvard Magazine* 123:42–47, 82–83.

Snyder-Beattie, A.E., Ord, T., and Bonsall, M.B. 2019. An upper bound for the background rate of human extinction. *Scientific Reports* 9:10554.

Thiault, L., et al. 2019. Escaping the perfect storm of simultaneous climate change impacts on agriculture and marine fisheries. *Science Advances* 5:eaaw9976.

Voosen, P. 2020. Project traces 500 million years of roller-coaster climate. *Science* 264:716–17.

Watts, R.G. 2007. *Global Warming and the Future of the Earth*. Williston, VT: Morgan and Claypool.

Wood, B., and Constantino, P. 2007. *Paranthropus boisei*: Fifty years of evidence and analysis. *Yearbook of Physical Anthropology* 50:106–32.

Index

Numbers in bold refer to pages with illustrations

www.ingramcontent.com/pod-product-compliance
Lightning Source LLC
Chambersburg PA
CBHW051955270326
41929CB00015B/2664